T0202709

Coriolis Vibratory Gyroscopes

Vladislav Apostolyuk

Coriolis Vibratory Gyroscopes

Theory and Design

 Springer

Vladislav Apostolyuk
Department of Aircraft Control Systems
and Instruments
National Technical University of Ukraine
Kiev
Ukraine

ISBN 978-3-319-35700-3   ISBN 978-3-319-22198-4   (eBook)
DOI 10.1007/978-3-319-22198-4

Springer Cham Heidelberg New York Dordrecht London

Printed on acid-free paper

Springer International Publishing AG Switzerland is part of Springer Science+Business Media
(www.springer.com)

Preface

Thanks to the advances in micromachining fabrication technologies and significant cost reduction due to mass production, miniature sensors of angular rate, or gyroscopes, found their way into the everyday life of every user of modern gadgets, such as smart phones, tablets or even wristwatches. Often without realising, many of us are carrying in our pockets fully equipped with all necessary sensors complete inertial navigation systems that not so long ago were available only for advanced vehicles in sea, land, air or space. Accelerometers and gyroscopes are found in specifications of any gadget supposed to react to user movements. And one of the most commonly used type of gyroscopes used to developed these systems is Coriolis vibratory gyroscope (CVG).

Needless to say that such a progress is a rewarding result of still ongoing research and development work of many scientists around the world. There are many books on design and fabrication of micromachined inertial sensors published by now, covering many important aspects of creating miniature inertial sensors. This book is intended to complete them with such aspects of gyroscopes development as theoretical analysis, mathematical modelling and analytical design, when desired performances are achieved by mathematical model analysis, rather than by trial and error in prototyping. Mathematical models and theoretical analysis are used both for optimal design of sensitive elements, as well as development of signal processing and control.

Chapter 1 gives an overview and classification of the most commonly used designs of CVG sensitive elements. Chapter 2 deals with deriving sensitive elements motion equations, and combining them into a single set of equations, covering most of the sensitive elements types. In Chap. 3 these generalised motions equations are solved for terms that can be used for angular rate measurements. Chapter 4 covers all aspects of CVG mathematical modelling in terms of demodulated envelope signals, thus providing the means for efficient analysis of sensitive elements dynamics as well as control and signal processing systems development. Calculation of major CVG performances is demonstrated in Chap. 5, along with an analytical approach to sensitive element design. The final Chap. 6 covers signal processing and control systems development. The latter Chaps. 5 and 6 are based on the results from previous chapters, and are the only two chapters that can be read independently of each other.

This book requires from the reader at least basic knowledge of higher mathematics, mechanics and control systems theory. So unfortunately, not everyone interested in vibratory gyroscope theory would be an intended reader, but rather graduate students and higher. At the same time, readers will not be required to follow the references and obtain additional information elsewhere as there are no references in this book. Many results presented in the book could be referenced to the previous publications of the author, but these references were intentionally omitted in order to relieve the reader from the burden of searching for important information on the subject elsewhere. Nevertheless, a time ordered list of recommended important publications on Coriolis vibratory gyroscopes is given at the end for those who may need it.

Finally, there is no finish line in research, development and writing books about them. So feel free to express your opinions, give suggestions and ask questions. Hope you will have plenty of them.

Contents

Chapter 1
Introduction

It is commonly accepted that the conventional gyroscope, as an instrument for determining orientation, was inspired by a top, known to practically all ancient civilisation. When a top rotates, it preserves its vertical orientation even if the base is tilted. This natural behaviour was used since eighteenth century A.D. to implement artificial horizon. The technical term *gyroscope* was introduced by Léon Foucault, when in 1852 he used a fast spinning rotor inside gimbals to demonstrate the Earth's rotation. The word *gyroscope* consists of two Greek words: *gyros*—"rotation" or "circle" and *scopeo*—"to see" or "observe". Although he did not invent the gyroscope itself, 1 year earlier, in 1851, he used a pendulum, known as *Foucault pendulum*, to demonstrate the Earth's rotation in another experiment. Curiously enough, this pendulum is commonly considered nowadays as a prototype to modern *Coriolis vibratory gyroscopes*.

1.1 Principle of Operation and Classification

Conventional mechanical gyroscope consists of a rotor mounted inside a two-frame gimbal that provides the rotor with two angular degrees of freedom in addition to one, corresponding to its own rotation. Such design is often referred to as a *free gyroscope*. Free gyroscope preserves orientation of axis of its own rotation relatively to the inertial reference frame, usually assumed to be fixed to stars. This phenomenon is used to determine orientation of the user relatively to the inertial frame when direct observation of orientation is impossible. When outer frame of a free gyroscope is removed and the remaining one provides only one angular degree of freedom, and when the base starts to rotate, the spinning rotor moves to align the axis of its own rotation with the axis of the external angular rate. This phenomenon is used to build such gyroscopic instruments as artificial horizons and gyroscopic compasses, where the external angular rate is provided by the Earth's rotation itself. However, if a spring is attached to the remaining gimbal frame, thus preventing the alignment between rotor and angular rate axes, the final deflection of the frame is linearly related to the magnitude of the external rate and allows its measurement. Such an instrument is called *angular rate sensor*. It turned out that using angular

© Springer International Publishing Switzerland 2016
V. Apostolyuk, *Coriolis Vibratory Gyroscopes*,
DOI 10.1007/978-3-319-22198-4_1

rate sensors instead of free gyroscopes is a cheaper, although less accurate, way to build inertial navigation instruments. Many different designs of angular rate sensors were developed since, and among them vibratory gyroscopes occupy niche of modern electro-mechanical sensors.

The main idea behind vibratory gyroscopes is to replace continuously spinning rotor with vibrating structure, and make use of the Coriolis effect, causing secondary motion that is related to the external angular rate. This type of angular rate sensors is commonly referred to as *Coriolis vibratory gyroscopes* (CVG).

In most of Coriolis vibratory gyroscopes, the sensitive element can be represented as an inertia element *m* and elastic suspension with two prevalent degrees of freedom: vertical and horizontal, as shown in Fig. 1.1.

Massive inertia element *m* is often called *proof mass*, similarly to the term used in accelerometers. The sensitive element is driven to oscillate at one of its modes with prescribed amplitude. This motion is usually called primary mode or primary motion. When the sensitive element rotates about a particular fixed-body axis, which is called sensitive axis (out-of-plane axis in Fig. 1.1), the so-called *Coriolis effect* causes the proof mass to move in an orthogonal direction, thus resulting in secondary motion. This effect is quantified by a fictitious Coriolis force

$$\vec{F}_C = -2 \cdot m \cdot \vec{\Omega} \times \vec{v}. \tag{1.1}$$

Here, $\vec{v}$ is the vector of the primary motion and $\vec{\Omega}$ is the external angular rate. In fact, we have two motions that are dynamically coupled via Coriolis effect. If there is no external angular rate ($\Omega = 0$), secondary motion is absent, since there is no Coriolis coupling.

In vibratory gyroscopes, this primary motion is designed to be oscillatory. As a result, secondary motion becomes oscillatory as well. Hereafter, excited oscillations are referred to as primary oscillations and oscillations caused by angular rate are referred to as secondary oscillations or secondary mode.

Fig. 1.1 Principle of CVG operation

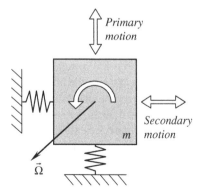

Contrary to the classical angular rate sensors based on gimballed electro-mechanical gyroscopes, information about external angular rate is contained in the amplitude of these secondary oscillations, rather than any gimbal deflection.

Utilising oscillatory motion instead of rotation as in conventional gyroscopes allows to avoid problems related to undesired motions caused by friction in gimbal bearing. It also significantly simplifies design of the sensitive element, since there are no electrical motors, which also makes vibratory gyroscopes amenable to miniaturisation.

It is also important to note that CVG can operate in two modes: as an angular rate sensor, measuring external angular rate, or in a so-called "whole-angle-mode", producing, in fact, angle of rotation, which is *integrated* angular rate. The latter can be achieved either by special sensitive element design, resulting in negligible damping, or by means of a specifically designed feedback control system. Ironically, although integrating CVG are considered as more advanced and complex compared to angular rate sensors, the very first vibratory gyroscope, namely Foucault pendulum, was an integrating CVG itself.

1.2 CVG Classification

In general, it is possible to design gyroscopes with different types of primary and secondary oscillations. For example, a combination of translation as primary oscillations and rotation as secondary oscillations was implemented in a so-called tuning fork gyroscope. It is worth mentioning that the nature of the primary motion does not necessarily have to be oscillatory but could be rotary as well. Such gyroscopes are called rotary vibratory gyroscopes. However, it is typically more convenient for the vibratory gyroscopes to be implemented with the same type and nature of primary and secondary oscillations.

With respect to the number of inertia elements used, the nature of primary and secondary motions of the sensitive element, classification of the vibratory gyroscopes can be represented as shown in Fig. 1.2.

Top-level separation is done based on the nature of the sensitive element motion. In our case, it can be either oscillatory or rotary. Classical dynamically tuned gyroscope (DTG) is an example of a rotary vibratory gyroscope.

Next level involves consideration of the general approach to design the sensitive element. In particular, design of the vibratory sensitive element can be based on continuous vibrating media or discrete (lumped masses). Corresponding mathematical models are based either on partial differential equations, namely modified wave equation, or systems of ordinary differential equations.

Third level affects the discrete masses branch of the classification and is based on a number of vibrating lumped masses (single or multiple).

Finally, CVG sensitive elements can be classified based on the combination of types of motions, utilised in primary and secondary motions: translational (linear) and rotational. In order to designate types of primary and secondary motions, the

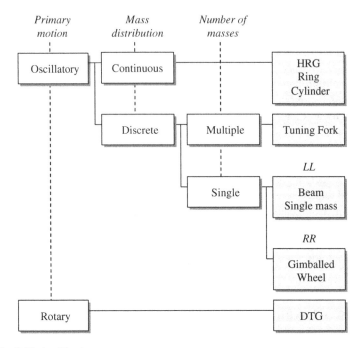

Fig. 1.2 CVG classification

following abbreviations are commonly accepted: "L" for linear and "R" for rotational. For example, if sensitive element utilises linear motion both for primary and secondary oscillations, it is designated as "LL".

The suggested here classification not only helps to apply proper mathematical modelling methods, but also serves as a guiding decision tree in the process of developing new types of CVG sensitive elements.

1.3 Sensitive Element Designs

Reasonable desire to miniaturise gyroscopes in order to expand the range of its applications resulted in utilising fabrication techniques from micro-electronics and micromachining to produce CVG sensitive elements. Such CVG elements are often referred to as micromechanical or MEMS (micro-electro-mechanical systems) gyroscopes. Because of quite strict limitations to the complexity of mechanical structures that can be produced using micromachining processes, the majority of the modern designs of micromechanical gyroscopes make use of a simple structure that consists of a single or multiple massive elements connected to the base by means of elastic suspension. The main purpose of the elastic suspension is to provide proof masses with at least two orthogonal degrees of freedom allowing primary and secondary oscillations. Another task, which is usually assigned to the design of elastic suspension, is to provide sufficient mechanical decoupling between primary and secondary oscillations, thus reducing so-called quadrature errors.

Let us now have a closer look at the specified in the classification above sensitive element designs. And we shall proceed not in the order of appearance in the classification, but with the most simple one.

1.3.1 Single Mass and Vibrating Beam

Conventional single-mass CVG sensitive element design originated from the very first vibratory gyroscope, namely Foucault pendulum. One could also soundly argue that the nature itself came out with this design; since owned by majority of flying insects, halters are just simple vibrating beam gyroscopes, allowing insects to control its flight.

There are few examples of such a single discrete mass gyroscope design. The first is a simple proof mass attached to the base by means of elastic stem that can deflect in two orthogonal directions (see Fig. 1.3).

Primary oscillations of such an inverted pendulum are excited along one direction, and when the base rotates around vertical axis, proof mass starts to oscillate in an orthogonal, secondary direction. As long as the CVG sensitive element consists of a single elastically suspended proof mass and no other essential masses are present in the design, such a design is referred to as a *single-mass* CVG.

In order to simplify fabrication process, the single-mass sensitive element can be produced in a form of square or triangular beams, as shown in Fig. 1.4.

Despite the fact that the beam itself has all the features of a continuous media vibratory sensor, its dynamics and operation as an angular rate sensor are

Fig. 1.3 Single-mass CVG
(*1* primary motion,
2 secondary motion)

Fig. 1.4 Vibrating beam
CVG (*1* primary motion,
2 secondary motion)

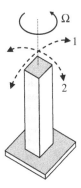

sufficiently well described in terms of ordinary differential equations. Therefore, this design has been placed in the discrete branch of the classification rather than distributed.

There are many different ways to drive and to detect motion of the beam-sensitive element. However, if the beam is made out of quartz, piezoelectric and piezoresistive phenomena are usually used to build driving and detecting systems.

Although vibrating beam is an easier to fabricate structure than inverted pendulum in Fig. 1.3, when micromachining technologies are involved, the preferred design for the sensitive element would be planar. If all structural components of a sensitive element are located in a single plane, it would be easier to fabricate them from a single silicon wafer, for instance. Example of such a planar single-mass design is shown in Fig. 1.5. Here, primary oscillations of the central proof mass *m* are excited usually by means of an electrostatic comb drive in an in-plane direction (1), and when the base starts to rotate, the secondary out-of-plane oscillations (2) are detected and used to measure the angular rate. Since both motions are translational, such a design would be labelled as "LL" (linear primary and linear secondary) according to the classification in Fig. 1.2.

Elastic suspension for the proof mass can be also designed allowing both primary and secondary motions to occur in a single plane (replicating schematics in Fig. 1.1). In this case, out-of-plane angular rate will be detected. Combining this

Fig. 1.5 Planar single-mass
CVG (*1* primary motion,
2 secondary motion)

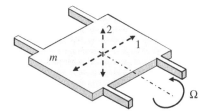

design with the one in Fig. 1.5, complete three orthogonal axis sensors set can be fabricated in a single planar silicon wafer.

1.3.2   Single Mass with Decoupling Frame

Simplicity of the single-mass CVG design, however, does not come without a price. Since a single set of flexible beams is used to facilitate both primary and secondary motions, even a small fabrication error, resulting in non-orthogonal elastic axes of the suspension beam, may cause a serious error, when secondary motion appears not due to the external angular rate, but is produced directly by the driving system via elastic cross-coupling. In order to overcome this problem, decoupling frame is added separating beams responsible for primary and secondary motions (Fig. 1.6). Proof mass m_1 is placed inside the decoupling frame m_2 and connected to it by means of flexible beams that allow the proof mass to move along the secondary direction only. The decoupling frame is fixed to the base via another set of beams that facilitate only primary motion for the decoupling frame. In case of both primary and secondary motions in-plane with the sensitive element, out-of-plane (vertical) angular rate is measured. Out-of-plane secondary motion for the proof mass can be provided by accordingly designed inner beams, and then in-plane angular rate can be measured.

Similar to the previous design, this sensitive element represents LL gyroscope too. However, if bending flexible beams of elastic suspension are replaced with the torsional elastic elements, thus providing rotational motion both for primary and secondary oscillations, this design can be transformed to RR gyroscope as shown in Fig. 1.7, and often referred to as *gimballed* CVG, since the decoupling frame resembles the gimbals from conventional gyroscopes. Here, both primary and

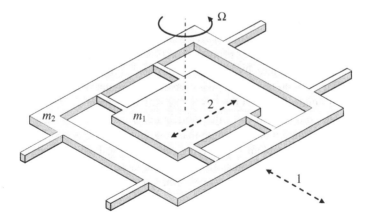

Fig. 1.6 Single-mass LL CVG with decoupling frame (*1* primary motion, *2* secondary motion)

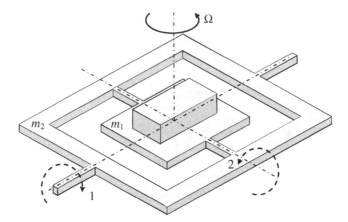

Fig. 1.7 Gimballed CVG (RR sensitive element) (*1* primary motion, *2* secondary motion)

secondary motions are rotational oscillations occurring around axes of corre-sponding torsional beams. Out-of-plane angular rate is measured when secondary rotational oscillations are detected. In order to create additional angular inertia properties for the central proof mass m_1, the massive inertia elements are usually added to it.

These additional inertia elements could have different shapes, such as brick (as shown in Fig. 1.7), sphere, cylinder, etc. The choice of the shape is driven by the desire to provide certain optimal relationships between moments of inertia for the sensitive element, but also depends on available fabrication technology.

1.3.3   Wheel Gyroscope

Another CVG sensitive element design that is based on rotational both primary and secondary motions is the so-called *wheel* gyroscope (see Fig. 1.8).

Massive disc is driven to primary rotational oscillations around vertical, out-of-plane axis. Two mutually orthogonal in-plane components Ω_x and Ω_y of the external angular rate can be sensed when secondary angular oscillations of the disc are detected around corresponding axes. Sensitive element of the wheel CVG represents more complicated from the fabrication point of view structure, if com-pared with the considered earlier single mass and gimballed gyroscopes. Designers are presented with the problem of providing capability for the disc to vibrate around three orthogonal axes. What comes in hand is the close resemblance of the wheel CVG with dynamically tuned gyroscopes, where rotor suspension structure solves essentially the same problem.

Fig. 1.8 Gimballed CVG
(RR sensitive element)
(*1* primary motion,
2 secondary motion)

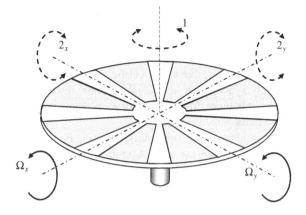

1.3.4 Tuning Fork Gyroscopes

After we have considered branches of single-mass gyroscopes according to the classification in Fig. 1.2, let us now move on to the branch of multiple proof masses. Classical tuning fork design is shown in Fig. 1.9.

Here, tines are driven to oscillate in the opposite directions, but along the same axis. When external rotation around vertical axis is applied, the secondary motion of tines occurs in opposite directions as well. In addition to this, Coriolis forces at each tine, being combined, produce harmonic torque around vertical axis. As a result, if the root stem allows rotation, the whole structure may as well start to oscillate around vertical axis.

Depending on the principle of the secondary motion detection, either secondary deflection of tines or rotation of the whole structure may be used to measure external angular rate.

The reason to have two, or more, proof masses instead of one is to be able to implement differential measurement scheme, when signals from detected secondary

Fig. 1.9 Conventional tuning fork (*dashed arrows* primary motion, *dotted* secondary motion)

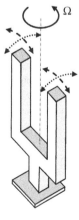

motions of each tine are combined (usually simply subtracted) in order to eliminate additive components in these signals. At the same time, useful part of the signals will be doubled. One could note that the same differential measurement could be implemented using two single-mass CVG, which are simply driven in opposite phases. However, due to the elastic coupling via common root stem, the tines will naturally synchronise its mutual motions, resulting in perfect counter-phase oscillations. It is more difficult to achieve the same tuning for two different, not coupled, single-mass sensitive elements.

To produce more possibilities to beneficially combine signals from different sources and make use of self-tuning, even more tines are added to the sensitive element (see Figs. 1.10 and 1.11). Additional tines also amplify the resulting Coriolis effect and allow more different modes of driving–sensing combinations as well.

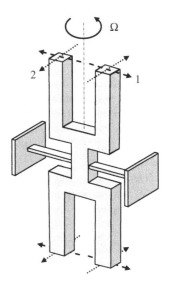

Fig. 1.10 Double tuning fork (*dashed arrows* primary motion, *dotted* secondary motion)

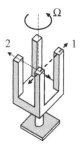

Fig. 1.11 Multiple tines tuning fork (*dashed arrows* primary motion, *dotted* secondary motion)

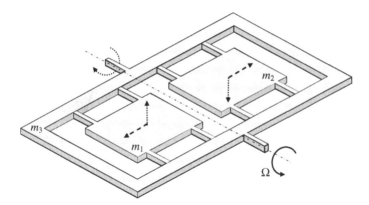

Fig. 1.12 Planar tuning fork CVG design (*dashed* primary motion, *dotted* secondary motion)

Similar to the simple vibrating beam, such tuning forks are usually fabricated from quartz allowing piezoelectric effect to be used for driving and sensing. However, if fabrication technologies from micro-electronics are to be utilised, then these designs have certain limitations due to the relative complexity of its three-dimensional structure. Certainly, simple planar design would preferable from the micromachining point of view. One of the examples of such designs is shown in Fig. 1.12.

Here, two proof masses m_1 and m_2 are driven to oscillate in opposite directions, and secondary motion, which is rotation of the common frame m_3 around sensitive axis (axis of the external angular rate), is detected in out-of-plane direction. Flexible beams of the proof masses are designed in such a way that they allow only primary motion of the proof masses to occur.

Since primary motion of the proof masses is translational and secondary is a rotation of the common frame, this design would be designated as LR according to our classification.

1.3.5 Hemispherical Resonating Gyroscope

One of the most well-known examples of the oscillatory gyroscope with continuous vibrating media is a hemispherical resonating gyroscope (HRG). HRG sensitive element design usually is based on the resonating shell that has a hemispheric or so-called "wine-glass" shape (see Fig. 1.13).

Primary oscillations are provided by a standing wave excited in the rim of the shell. In case of no external angular rate, there is no motion at nodes of the wave, which are located at 45° from the primary axis 1. If the sensitive element rotates around sensitive axis, which is orthogonal to the plane of the wave, Coriolis effect causes the wave to shift around the rim. As a result, oscillations can be now

Fig. 1.13 Hemispherical
resonating gyroscope
(*dashed* primary motion,
dotted secondary motion)

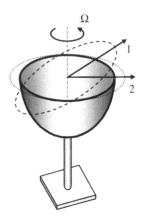

detected at the nodes, and these secondary oscillations are related to the external
angular rate. Despite high performances provided by the conventional HRG,
complexity and high cost of fabricating perfect vibrating hemisphere prohibits its
massive usage, including civil applications. In order to reduce the cost along with
acceptable degradation of performances, several simplified designs were introduced. In particular, the hemispherical shape of the shell has been replaced with a
thin cylinder (Fig. 1.14) or ring (Fig. 1.15). All these designs are aimed at providing
vibrating capability for the ring-shaped continuous medium with possibility to drive
and detect oscillations in this ring.

The latter ring-shaped design is also suitable for applying micromachining and
has certain potential for miniaturisation. The major shortcoming of this simplifi-
cation comes from the imperfections of the resonator. As a result, comparatively
high damping and secondary oscillations that are not caused by the external angular
rate significantly reduce performances of these gyroscopes.

However, there are plenty of not so-demanding applications where these low
grade sensors can be successfully used.

Fig. 1.14 Cylindrical CVG
(*dashed* primary motion,
dotted secondary motion)

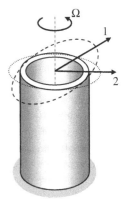

Fig. 1.15 Ring-shaped CVG
(*dashed* primary motion,
dotted secondary motion)

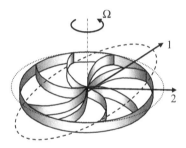

Resume

The difference between most of all modern Coriolis vibratory gyroscopes lays mainly in designs of the mass itself and the elastic suspensions rather than operation principle and its mathematical model. Needless to say that such difference is driven almost solely by specific features of chosen fabrication process. Therefore, mathematical models and design methodologies presented later in this book can be directly applied to the analysis of all Coriolis vibratory gyroscopes. In case of continuous media sensors, the results still can be applied to a certain extent, provided lumped mass mathematical representation is used.

Chapter 2
Motion Equations of Coriolis Vibratory Gyroscopes

The first and one of the most important steps in analysis of any mechanical system is to derive its motion equations. Such equations, being solved either numerically or in a closed form, allow all forms of dynamics analysis and design optimisation to be performed on the system under consideration. The central mechanical part of any Coriolis vibratory gyroscope is its sensitive element. No matter what specific design has been utilised, the sensitive element must be capable of providing at least two forms of mechanical oscillations: primary and one or more secondary. While the former is intentionally induced into the mechanical structure, the latter will appear due to the Coriolis force when the sensitive element is rotated.

In this chapter, we shall derive differential motion equations for sensitive elements of all commonly used designs and then generalise them to obtain a single set of equations suitable for different CVG designs.

2.1 Translational Sensitive Element Motion Equations

Translational motion sensitive element utilises translational motion for both primary and secondary oscillations. It is also commonly accepted to refer to the translational sensitive element as an LL-gyro (linear primary and linear secondary motion).

Figure 2.1 shows the schematic representation of the structure implementing translational type of kinematics. In the most generalised form, sensitive element consists of the proof mass (m_2), the decoupling frame (m_1) and two sets of elastic elements ("springs") connecting masses to each other and to the base. In addition to that, let us introduce the right-handed orthogonal and normalised reference frame $OXYZ$, in which primary oscillations are excited along the axis Y, then the secondary oscillations occur along the axis X and therefore the third axis Z is considered as a sensitive axis. The latter means that in an ideal case, external rotation around this axis will be sensed by the sensitive element.

Let us define position $\vec{X}_1$ of the decoupling frame and position $\vec{X}_2$ of the proof mass in the reference frame $OXYZ$ as

© Springer International Publishing Switzerland 2016
V. Apostolyuk, *Coriolis Vibratory Gyroscopes*,
DOI 10.1007/978-3-319-22198-4_2

Fig. 2.1 Translational
sensitive element of CVG

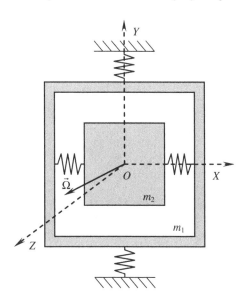

$$\vec{X}_1 = \{0, x_1, 0\},$$
$$\vec{X}_2 = \{x_2, x_1, 0\}. \tag{2.1}$$

Here x_1 is the displacement of the decoupling frame relatively to the fixed base, and x_2 is the displacement of the proof mass relatively to the decoupling frame. Hereinafter, subscripts 1 and 2 stand for primary and secondary motions of the sensitive element and should not be confused with the axis number.

For the sake of generality, let us assume that the sensitive element rotation is given by an arbitrary angular rate vector $\vec{\Omega}$, which is defined by its projections on the introduced above reference frame as $\vec{\Omega} = \{\Omega_x, \Omega_y, \Omega_z\}$.

In order to derive motion equations of the sensitive element, let us use the Lagrange equation in the following form:

$$\frac{d}{dt}\left(\frac{\partial L}{\partial \dot{x}_i}\right) - \frac{\partial L}{\partial x_i} = Q_i. \tag{2.2}$$

Here $L = E_k - E_p$ is the Lagrange's function, E_k and E_p are the total kinetic and potential energies of the sensitive element, Q_i is the generalised force acting on the sensitive element, and subscript i ranges from 1 for the primary motion to the number secondary motions under consideration.

So to use the Lagrange Eq. (2.2), proper expressions for the kinetic and potential energies for the sensitive element must be obtained.

In the most generalised case, kinetic energy of a moving point mass m is

$$E_k = \frac{m}{2}(\vec{V} \cdot \vec{V}), \qquad (2.3)$$

where $\vec{v}$ is the absolute velocity vector that can be expressed in terms of the point mass position $\vec{X}$ and angular rate $\vec{\Omega}$ of the reference frame rotation as

$$\vec{V} = \dot{\vec{X}} = \frac{d}{dt}\vec{X} + \vec{\Omega} \times \vec{X}. \qquad (2.4)$$

Here $\frac{\tilde{d}}{dt}\vec{X} = \{\dot{X}_x, \dot{X}_y, \dot{X}_z\}$ is the local derivative of the vector $\vec{X}$ in the rotating frame. Velocities vectors for the decoupling frame and the proof mass positions (2.1) calculated using (2.4) are

$$\begin{aligned}
\vec{V}_1 &= \{-x_1\Omega_z, \dot{x}_1, x_1\Omega_x\}, \\
\vec{V}_2 &= \{-x_1\Omega_z + \dot{x}_2, x_2\Omega_z + \dot{x}_1, x_1\Omega_x - x_2\Omega_y\}.
\end{aligned} \qquad (2.5)$$

Corresponding to (2.5), kinetic energies for primary and secondary motions are as follows:

$$\begin{aligned}
E_{k1} &= \frac{m_1}{2}[\dot{x}_1^2 + x_1^2(\Omega_x^2 + \Omega_z^2)], \\
E_{k2} &= \frac{m_2}{2}[(\dot{x}_1 + x_2\Omega_z)^2 + (\dot{x}_2 - x_1\Omega_z)^2 + (x_1\Omega_x - x_2\Omega_y)^2], \\
E_k &= E_{k1} + E_{k2}.
\end{aligned} \qquad (2.6)$$

Here E_k is the total kinetic energy of the CVG sensitive element as a sum of the respective kinetic energies of the decoupling frame and the proof mass.

The second term in the Lagrange function is the total potential energy of springs in the elastic suspension of the sensitive element

$$E_p = \frac{k_1}{2}x_1^2 + \frac{k_2}{2}x_2^2, \qquad (2.7)$$

where k_1 is the total stiffness of the elastic suspension along the axis Y (primary motion) and k_2 is the total stiffness along the axis X (secondary motion).

Combining expressions (2.6) and (2.7) into the Lagrange function and using the result with the Lagrange Eq. (2.2) gives us the following system of two differential equations, describing the motion of the CVG sensitive element:

$$\begin{cases}
(m_1 + m_2)\ddot{x}_1 + [k_1 - (m_1 + m_2)(\Omega_x^2 + \Omega_z^2)]x_1 + m_2((\Omega_x\Omega_y + \dot{\Omega}_z)x_2 + 2\Omega_z\dot{x}_2) = Q_1, \\
m_2\ddot{x}_2 + [k_2 - m_2(\Omega_y^2 + \Omega_z^2)]x_2 + m_2(-2\Omega_z\dot{x}_1 + (\Omega_x\Omega_y - \dot{\Omega}_z)x_1) = Q_2.
\end{cases}$$

These equations can now be rewritten by dividing both parts of the equations by its corresponding higher order derivative terms coefficients ($m_1 + m_2$ for the first equation and m_2 for the second). The result is

$$\begin{cases} \ddot{x}_1 + (\omega_1^2 - \Omega_x^2 - \Omega_z^2)x_1 + 2d\Omega_z\dot{x}_2 + d(\Omega_x\Omega_y + \dot{\Omega}_z)x_2 = q_1, \\ \ddot{x}_2 + (\omega_2^2 - \Omega_y^2 - \Omega_z^2)x_2 - 2\Omega_z\dot{x}_1 + (\Omega_x\Omega_y - \dot{\Omega}_z)x_1 = q_2, \end{cases} \quad (2.8)$$

where $\omega_1^2 = k_1/(m_1 + m_2)$ and $\omega_2^2 = k_2/m_2$ are the squared natural frequencies of the primary and secondary motions, respectively, $d = m_2/(m_1 + m_2)$ is the dimensionless inertia asymmetry factor, $q_1 = Q_1/(m_1 + m_2)$, $q_2 = Q_2/m_2$ are the generalised accelerations from different external forces that act along respective axes.

Equation (2.8) describes the motion of the generalised CVG with translational sensitive element. Note that if we simply assume that the mass of the decoupling frame is zero ($m_1 = 0$), then $d = 1$, and we can obtain motion equations for the single-mass CVG without decoupling frame, which, for example, corresponds to a vibrating beam design.

Finally, Eq. (2.8) must be completed by adding damping forces terms, producing translational sensitive element motion equations

$$\begin{cases} \ddot{x}_1 + 2\zeta_1\omega_1\dot{x}_1 + (\omega_1^2 - \Omega_x^2 - \Omega_z^2)x_1 + 2d\Omega_z\dot{x}_2 + d(\Omega_x\Omega_y + \dot{\Omega}_z)x_2 = q_1, \\ \ddot{x}_2 + 2\zeta_2\omega_2\dot{x}_2 + (\omega_2^2 - \Omega_y^2 - \Omega_z^2)x_2 - 2\Omega_z\dot{x}_1 + (\Omega_x\Omega_y - \dot{\Omega}_z)x_1 = q_2. \end{cases} \quad (2.9)$$

Here ζ_1 and ζ_2 are the dimensionless damping factors that correspond to the primary and secondary motions of the sensitive element.

Equation (2.9) contain all components of the angular rate vector $\vec{\Omega}$, but only component Ω_z is present as a first-order angular rate term and therefore will be properly measured by translational CVG. Let us now assume that the angular rate coincides with the axis Z, which means $\vec{\Omega} = \{0, 0, \Omega\}$. This assumption leads to the simpler form of motion equations as

$$\begin{cases} \ddot{x}_1 + 2\zeta_1\omega_1\dot{x}_1 + (\omega_1^2 - \Omega^2)x_1 = q_1 - 2d\Omega\dot{x}_2 - d\dot{\Omega}x_2, \\ \ddot{x}_2 + 2\zeta_2\omega_2\dot{x}_2 + (\omega_2^2 - \Omega^2)x_2 = q_2 + 2\Omega\dot{x}_1 + \dot{\Omega}x_1. \end{cases} \quad (2.10)$$

Taking a closer look at (2.10), one can see that in case of an ideal elastic suspension, where there is no cross-coupling between the decoupling frame and the proof mass, primary and secondary motions in the system are coupled only by means of the angular rate Ω. It means that given the absence of any external forces acting on the proof mass along generalised coordinate x_2 (e.g. $q_2 = 0$), any forced displacements in this direction will be caused by the angular rate alone. At the same time, angular rate in the Eq. (2.10) is unknown and, generally speaking, varies in time. This means that although the obtained system of sensitive element motion equations is linear, it contains unknown time-dependent coefficients. Needless to say that it is quite a complicated task to find a closed-form analytical solution for such system. However, the good news is that in order to analyse CVG dynamics and being able to optimise its performances, we do not need to solve these equations with respect to the sensitive element displacements x_1 and x_2.

2.2   Rotational Sensitive Element Motion Equations

Contrary to the translational sensitive element, rotational design utilises rotation for both primary and secondary oscillations. For this reason, it is often referred to as a RR-gyro (rotational primary and rotational secondary motion).

Similar to the translational sensitive element, Fig. 2.2 demonstrates rotational CVG sensitive element kinematics.

Linear (translational) springs are replaced with the rotational elastic torsions, and generalised coordinates now represent angles rather than translational displacements. Here α_1 corresponds to the angle between base and the decoupling frame, and α_2 corresponds to the angle between decoupling frame and the proof mass element. These angles are commonly referred to as Euler angles. Transformations of the reference frame axes after each of these rotations (primary and secondary) are demonstrated in Fig. 2.3.

Here reference frame $OX_1Y_1Z_1$ is the result of rotating the reference frame $OXYZ$ around Y axis by the angle α_1, and is fixed to the decoupling frame. Similarly, reference frame $OX_2Y_2Z_2$ is obtained by rotating frame $OX_1Y_1Z_1$ by the angle α_2, and is fixed to the proof mass element.

Fig. 2.2 Rotational sensitive element of CVG

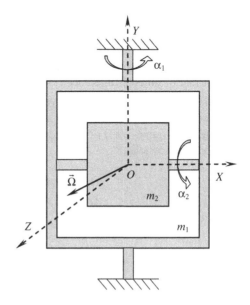

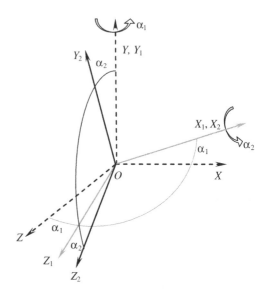

Fig. 2.3 Axes
transformations due to the
primary and secondary
motion

In order to derive rotational CVG sensitive element motion equation, we need to
start again with the proper expressions for kinetic and potential energies.

The difference with the previous case is that the motion is rotational now, and the
corresponding kinetic energy is expressed in terms of angular rate instead of
velocities and moments of inertia instead of mass:

$$E_k = \frac{I}{2}\Omega^2. \tag{2.11}$$

Here I is the moment of inertia of the rotating element around axis, which
coincides with the direction of the angular rate vector $\vec{\Omega}$. Apparently, this case is
slightly more complicated comparing to the translational motion of the CVG sensitive element, since proper moments of inertia are applied to components of the
angular rate in different reference frames.

Initially, external angular rate vector $\vec{\Omega}$ is defined by its components $\vec{\Omega} =$
$\{\Omega_x, \Omega_y, \Omega_z\}$ in the reference frame $OXYZ$. Transformation of its components into
the reference frame $OX_1Y_1Z_1$, fixed to the decoupling frame, are given by the
following expressions:

$$\begin{aligned}
\Omega_{x1} &= \Omega_x \cos \alpha_1 - \Omega_z \sin \alpha_1, \\
\Omega_{y1} &= \Omega_y + \dot{\alpha}_1, \\
\Omega_{z1} &= \Omega_x \sin \alpha_1 + \Omega_z \cos \alpha_1.
\end{aligned} \tag{2.12}$$

Further transformation of the angular rate into the reference frame $OX_2Y_2Z_2$
assigned to the proof mass element is performed in a similar way:

$$\Omega_{x2} = \Omega_{x1} + \dot{\alpha}_2,$$
$$\Omega_{y2} = \Omega_{y1} \cos \alpha_2 + \Omega_{z1} \sin \alpha_2 \qquad (2.13)$$
$$\Omega_{z2} = -\Omega_{y1} \sin \alpha_2 + \Omega_{z1} \cos \alpha_2.$$

Kinetic energy (2.11) can now be expressed in terms of different angular rates components (2.12) and (2.13) as

$$E_k = \frac{1}{2} \left(I_{1x}\Omega_{x1}^2 + I_{1y}\Omega_{y1}^2 + I_{1z}\Omega_{z1}^2 + I_{2x}\Omega_{x2}^2 + I_{2y}\Omega_{y2}^2 + I_{2z}\Omega_{z2}^2 \right). \qquad (2.14)$$

Here I_{ix}, I_{iy}, I_{iz} are the moments of inertia of the ith element ($i = 1$ corresponds to the decoupling frame, $i = 2$ corresponds to the proof mass element) around respective axis of the corresponding reference frame.

Potential energy of the sensitive element is similar to the case of translational motions except for the angular stiffness of the torsions instead of the springs:

$$E_p = \frac{k_1}{2} \alpha_1^2 + \frac{k_2}{2} \alpha_2^2. \qquad (2.15)$$

Here k_i are the angular spring constants of the elastic suspension.

Substituting obtained expressions for the kinetic (2.14) and potential (2.15) energies in the Lagrange Eq. (2.2) results in rather big equations that are difficult to analyse. At the same time, these equations are nonlinear due to the presence of sine and cosine functions of the generalised variables α_1 and α_2.

However, taking into consideration that elastic suspensions usually do not allow big angular deflections, it is reasonable to assume that angles α_1 and α_2 are small. Hence, obtained motion equations can be linearized using Taylor series expansion that results in the following linear in terms of angles equations:

$$\begin{cases} (I_{1y} + I_{2y})(\ddot{\alpha}_1 + \dot{\Omega}_y) + [k_1 + (I_{1x} - I_{1z} + I_{2x} - I_{2z})(\Omega_x^2 - \Omega_z^2)]\alpha_1 \\ \quad + (I_{2x} + I_{2y} - I_{2z})\Omega_z\dot{\alpha}_2 - (I_{2y} - I_{2z})(\Omega_x\Omega_y - \dot{\Omega}_z)\alpha_2 \\ \quad + (I_{1x} - I_{1z} + I_{2x} - I_{2z})\Omega_x\Omega_z = Q_1, \\ I_{2x}(\ddot{\alpha}_2 + \dot{\Omega}_x) + [k_2 + (I_{2y} - I_{2z})(\Omega_y^2 - \Omega_z^2)]\alpha_2 \\ \quad - (I_{2x} + I_{2y} - I_{2z})\Omega_z\dot{\alpha}_1 + (I_{2y} - I_{2z})\Omega_x\Omega_y\alpha_1 - I_{2x}\dot{\Omega}_z\alpha_1 \\ \quad - (I_{2y} - I_{2z})\Omega_y\Omega_z = Q_2. \end{cases} \qquad (2.16)$$

Here Q_1 and Q_2 are the generalised torques acting around primary and secondary rotations. Let us now divide both sides of the equations by the corresponding highest derivative coefficients:

$$
\begin{cases}
\ddot{\alpha}_1 + \dot{\Omega}_y + \left[\dfrac{k_1}{I_{1y}+I_{2y}} + \dfrac{I_{1x}-I_{1z}+I_{2x}-I_{2z}}{I_{1y}+I_{2y}}(\Omega_x^2-\Omega_z^2) \right]\alpha_1 \\[2ex]
\quad + \dfrac{I_{2x}+I_{2y}-I_{2z}}{I_{1y}+I_{2y}}\Omega_z\dot{\alpha}_2 - \dfrac{I_{2y}-I_{2z}}{I_{1y}+I_{2y}}(\Omega_x\Omega_y-\dot{\Omega}_z)\alpha_2 \\[2ex]
\quad + \dfrac{I_{1x}-I_{1z}+I_{2x}-I_{2z}}{I_{1y}+I_{2y}}\Omega_x\Omega_z = \dfrac{Q_1}{I_{1y}+I_{2y}}, \\[2ex]
\ddot{\alpha}_2 + \dot{\Omega}_x + \left[\dfrac{k_2}{I_{2x}} + \dfrac{I_{2y}-I_{2z}}{I_{2x}}(\Omega_y^2-\Omega_z^2) \right]\alpha_2 \\[2ex]
\quad - \dfrac{I_{2x}+I_{2y}-I_{2z}}{I_{2x}}\Omega_z\dot{\alpha}_1 + \dfrac{I_{2y}-I_{2z}}{I_{2x}}\Omega_x\Omega_y\alpha_1 - \dot{\Omega}_z\alpha_1 \\[2ex]
\quad - \dfrac{I_{2y}-I_{2z}}{I_{2x}}\Omega_y\Omega_z = \dfrac{Q_2}{I_{2x}}.
\end{cases} \tag{2.17}
$$

The following new variables can now be introduced in order to simplify Eq. (2.17):

$q_1 = Q_1/(I_{1y}+I_{2y})$, $q_2 = Q_2/I_{2x}$—are the generalised angular accelerations from the corresponding external torques, $\omega_1^2 = k_1/(I_{1y}+I_{2y})$ and $\omega_2^2 = k_2/I_{2x}$—are the squared natural frequencies of the primary and secondary motions correspondingly, $g_1 = (I_{2x}+I_{2y}-I_{2z})/(I_{1y}+I_{2y})$ and $g_2 = (I_{2x}+I_{2y}-I_{2z})/I_{2x}$—are the gyroscopic Coriolis coefficients, $d_1 = (I_{1x}-I_{1z}+I_{2x}-I_{2z})/(I_{1y}+I_{2y})$, $d_2 = (I_{2y}-I_{2z})/I_{2x}$, $d_3 = (I_{2y}-I_{2z})/(I_{1y}+I_{2y})$—are the coefficients that can be seen as sensitive element design parameters along with the gyroscopic coefficients. All these coefficients allow to rewrite Eq. (2.17) as follows:

$$
\begin{cases}
\ddot{\alpha}_1 + 2\zeta_1\omega_1\dot{\alpha}_1 + [\omega_1^2 + d_1(\Omega_x^2-\Omega_z^2)]\alpha_1 + g_1\Omega_z\dot{\alpha}_2 \\[1ex]
\quad - d_3(\Omega_x\Omega_y-\dot{\Omega}_z)\alpha_2 + d_1\Omega_x\Omega_z + \dot{\Omega}_y = q_1, \\[1ex]
\ddot{\alpha}_2 + 2\zeta_2\omega_2\dot{\alpha}_2 + [\omega_2^2 + d_2(\Omega_y^2-\Omega_z^2)]\alpha_2 - g_2\Omega_z\dot{\alpha}_1 \\[1ex]
\quad + d_2\Omega_x\Omega_y\alpha_1 - \dot{\Omega}_z\alpha_1 - d_2\Omega_y\Omega_z + \dot{\Omega}_x = q_2.
\end{cases} \tag{2.18}
$$

Here damping terms were also added that are characterised by the dimensionless damping factors ζ_1 and ζ_2. Finally, similarly to the case of translational CVG sensitive element, we can assume that angular rate coincides with the axis Z (e.g. $\vec{\Omega} = \{0,0,\Omega\}$). This gives us the much simpler form of the Eq. (2.18):

$$
\begin{cases}
\ddot{\alpha}_1 + 2\zeta_1\omega_1\dot{\alpha}_1 + (\omega_1^2 - d_1\Omega^2)\alpha_1 = q_1 - g_1\Omega\dot{\alpha}_2 - d_3\dot{\Omega}\alpha_2, \\[1ex]
\ddot{\alpha}_2 + 2\zeta_2\omega_2\dot{\alpha}_2 + (\omega_2^2 - d_2\Omega^2)\alpha_2 = q_2 + g_2\Omega\dot{\alpha}_1 + \dot{\Omega}\alpha_1.
\end{cases} \tag{2.19}
$$

Rotational CVG sensitive element motion equations in the form (2.19) have exactly the same structure as the Eq. (2.10) for the translational CVG. The only difference is in the coefficients and meaning of the generalised coordinates, which describe either translational or rotational motion. This fact will allow us later to produce a single set of CVG sensitive element motion equations, leading to unified analysis of both types of designs.

2.3 Tuning Fork Sensitive Element Motion Equations

We have already derived motion equations for CVG sensitive elements utilising either translational or rotational motion for both primary and secondary modes. Let us now move on to the design that uses combination of translational and rotational motions, namely "tuning fork" CVG. Kinematics scheme of the tuning fork sensitive element is shown in Fig. 2.4.

Tuning fork sensitive element consists of two identical proof masses m_1 and m_2 attached using linear springs to the common frame m_3, which is attached to the base by means of elastic torsions. Proof masses are able to move along axis Y, and when the sensitive element rotates around axis Z, due to the Coriolis force the whole sensitive element starts to rotate around axis Z (angle α). For the sake of simplicity, it is also assumed that the external angular rate is perfectly aligned with the axis Z ($\vec{\Omega} = \{0, 0, \Omega\}$. Position of the proof masses in the reference frame fixed to the common frame are given by the following two vectors:

$$\vec{X}_1 = \{0, -r + y_1, 0\},$$
$$\vec{X}_2 = \{0, r + y_2, 0\}. \tag{2.20}$$

Fig. 2.4 Tuning fork sensitive element

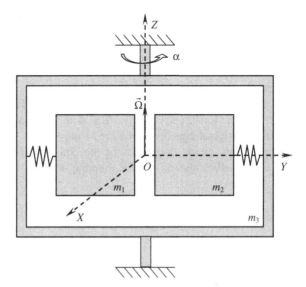

Here r is the constant distance from the axis Z to the centres of the proof masses when they are not moving, and the motion is described by the displacements y_1 and y_2. Total angular rate of the frame rotation is

$$\vec{\Omega}_3 = \{0, 0, \Omega + \dot{\alpha}\}. \tag{2.21}$$

Corresponding to (2.20) velocity vectors are as follows:

$$\begin{aligned}
\vec{V}_1 &= \{(r - y_1)(\Omega + \dot{\alpha}), \dot{y}_1, 0\}, \\
\vec{V}_2 &= \{-(r + y_2)(\Omega + \dot{\alpha}), \dot{y}_2, 0\}.
\end{aligned} \tag{2.22}$$

Total kinetic energy of the sensitive element obtained using velocities (2.22) for the expression (2.3) (translational motion) and angular rate (2.21) for the expression (2.11) (rotational motion) is

$$\begin{aligned}
E_k = &\frac{m_1}{2}\left[(r - y_1)^2(\Omega + \dot{\alpha})^2 + \dot{y}_1^2\right] + \frac{m_2}{2}\left[(r + y_2)^2(\Omega + \dot{\alpha})^2 + \dot{y}_2^2\right] \\
&+ \frac{I_{3z}}{2}(\Omega + \dot{\alpha})^2.
\end{aligned} \tag{2.23}$$

Here I_{3z} is the moment of inertia of the common frame around axis Z. Potential energy of the elastic suspension is

$$E_p = \frac{k_1}{2}y_1^2 + \frac{k_2}{2}y_2^2 + \frac{k_3}{2}\alpha^2. \tag{2.24}$$

Finally, substituting energies (2.23) and (2.24) into the Lagrange Eq. (2.2) results in the following tuning fork sensitive element motion equations:

$$\begin{cases}
m_1\ddot{y}_1 + (k_1 - m_1\Omega^2)y_1 + 2m_1 r\Omega\dot{\alpha} + m_1 r\Omega^2 = Q_1, \\
m_2\ddot{y}_2 + (k_2 - m_2\Omega^2)y_2 - 2m_2 r\Omega\dot{\alpha} - m_2 r\Omega^2 = Q_2, \\
I_z\ddot{\alpha} + k_3\alpha - 2r(m_1\dot{y}_1 - m_2\dot{y}_2)\Omega - 2r(m_1 y_1 - m_2 y_2)\dot{\Omega} + I_z\dot{\Omega} = Q_3.
\end{cases} \tag{2.25}$$

Here $I_z = I_{3z} + (m_1 + m_2)r^2$ is the total moment of inertia of the sensitive element around Z axis, Q_i are the generalised forces for translational variables and torque for rotational, respectively. Also note that motion Eq. (2.25) are the linear part of the original equations, obtained from (2.2).

Let us now divide both sides of the equations by the corresponding highest derivative coefficients and add the damping terms:

$$\begin{cases}
\ddot{y}_1 + (\omega_1^2 - \Omega^2)y_1 + 2r\Omega\dot{\alpha} + r\Omega^2 = q_1, \\
\ddot{y}_2 + (\omega_2^2 - \Omega^2)y_2 - 2r\Omega\dot{\alpha} - r\Omega^2 = q_2, \\
\ddot{\alpha} + \omega_3^2\alpha - 2\dfrac{r}{I_z}(m_1\dot{y}_1 - m_2\dot{y}_2)\Omega - 2\dfrac{r}{I_z}(m_1 y_1 - m_2 y_2)\dot{\Omega} + \dot{\Omega} = q_3.
\end{cases} \tag{2.26}$$

Here $\omega_1^2 = k_1/m_1$, $\omega_2^2 = k_2/m_2$ and $\omega_3^2 = k_3/I_z$ are the squared natural frequencies of the corresponding motions; $q_1 = Q_1/m_1$, $q_2 = Q_2/m_2$ and $q_3 = Q_3/I_z$ are the accelerations from the external forces, acting on the corresponding elements of the sensitive element.

System of Eq. (2.26) can be further simplified if the new variable $y = y_1 - y_2$ is introduced and proof masses along with its elastic suspensions are assumed to be identical ($m_1 = m_2 = m$, $k_1 = k_2 = k$). As a result, first two equations are reduced to one, and the tuning fork sensitive element motion equations become

$$\begin{cases} \ddot{y} + 2\zeta_y\omega_y\dot{y} + (\omega_y^2 - \Omega^2)y + 2r\Omega(2\dot{\alpha} + \Omega) = q_y, \\ \ddot{\alpha} + 2\zeta_3\omega_3\dot{\alpha} + \omega_3^2\alpha - 2m\dfrac{r}{I_z}(\Omega\dot{y} + y\dot{\Omega}) + \dot{\Omega} = q_3, \end{cases} \tag{2.27}$$

where $q_y = q_1 - q_2$, $\omega_y^2 = k/m$, ζ_y and ζ_3 are the dimensionless damping factors of the added damping terms.

Analysis of the motions Eq. (2.27) shows that provided masses are set to oscillate in the opposite phases, angular motion of the sensitive element becomes dependent on the external angular rate, which allows its measurement.

2.4 Ring-Shaped Sensitive Element Motion Equations

While all previously considered designs use lumped masses to sense Coriolis acceleration, distributed masses are also used and provide excellent performances in numerous applications. As an example of the distributed masses design, let us consider ring-shaped sensitive element of CVG, shown in Fig. 2.5.

Ring-shaped sensitive element is set to oscillate along axes X and Y (primary ring displacement x_1). These oscillations could be also viewed as a standing wave excited in the ring. When there is no external rotation applied to the sensitive

Fig. 2.5 Ring-shaped CVG sensitive element

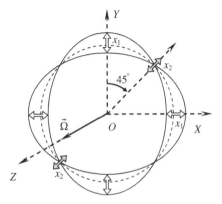

element, primary wave has four nodes, where the ring displacement is zero. These nodes are located at $45°$ between axes X and Y.

When the sensitive element rotates around axis Z, the primary wave starts to shift around the ring due to the Coriolis force. As a result, secondary oscillations (ring displacement x_2) start to appear in the nodes, thus allowing measurements of the external angular rate $\vec{\Omega}$.

Despite the fact that there are no lumped masses in the ring sensitive element, its motion equations in terms of the primary and secondary displacements x_1 and x_2 still could be written in a lumped form as

$$\begin{cases} \ddot{x}_1 + 2\zeta\omega\dot{x}_1 + (\omega^2 - \Omega^2)x_1 = q_1 - 2c\Omega\dot{x}_2 - c\dot{\Omega}x_2, \\ \ddot{x}_2 + 2\zeta\omega\dot{x}_2 + (\omega^2 - \Omega^2)x_2 = q_2 + 2c\Omega\dot{x}_1 + c\dot{\Omega}x_1. \end{cases} \tag{2.28}$$

Here ω is the natural frequency of the ring, ζ is the dimensionless damping factor, Ω is the projection of the external angular rate to the axis Z, c is the gyroscopic coupling factor (Brian coefficient), q_1 and q_2 are the effective accelerations from external forces acting along primary and secondary displacements, respectively. Apparently, motion Eq. (2.28) are quite similar to the translational sensitive element equations in case when there is no decoupling frame.

2.5 Generalised Motion Equations

It is no coincidence that motion equations of all considered above sensitive element designs look so similar, since they are all based on the same principle—Coriolis acceleration measurement by vibrating structures. Let us now write down CVG sensitive element motion equations in a form that is applicable to all designs:

$$\begin{cases} \ddot{x}_1 + 2\zeta_1\omega_1\dot{x}_1 + (\omega_1^2 - d_1\Omega^2)x_1 = q_1 - g_1\Omega\dot{x}_2 - d_3\dot{\Omega}x_2, \\ \ddot{x}_2 + 2\zeta_2\omega_2\dot{x}_2 + (\omega_2^2 - d_2\Omega^2)x_2 = q_2 + g_2\Omega\dot{x}_1 + d_4\dot{\Omega}x_1. \end{cases} \tag{2.29}$$

Here x_1 and x_2 are the generalised displacements describing primary and secondary motion of the sensitive element, either translational or rotational; ω_1 and ω_2 are the natural frequencies, ζ_1 and ζ_2 are the damping factors of the primary and secondary motions correspondingly; q_1 and q_2 are the accelerations (either translational or rotational) from external forces/torques, Ω is the external angular rate, orthogonal to the primary and secondary motions. Remaining coefficients are the functions of the sensitive element design parameters and are given in the Table 2.1.

Note that in order to make tuning fork motion Eq. (2.27) compatible with the generalised form (2.29), certain terms were dropped. However, these modifications do not make the generalised form less applicable to analysis of the tuning fork

Table 2.1 Design-dependent generalised equations coefficients

Design:	Beam	LL-gyro	RR-gyro	Tuning fork	Ring
g_1	2	$\frac{2m_2}{m_1+m_2}$	$\frac{I_{2x}+I_{2y}-I_{2z}}{I_{1y}+I_{2y}}$	$4r$	$2c$
g_2	2	2	$\frac{I_{2x}+I_{2y}-I_{2z}}{I_{2x}}$	$\frac{2mr}{I_z}$	$2c$
d_1	1	1	$\frac{I_{1x}-I_{1z}+I_{2x}-I_{2z}}{I_{1y}+I_{2y}}$	1	1
d_2	1	1	$\frac{I_{2y}-I_{2z}}{I_{2x}}$	0	1
d_3	1	$\frac{m_2}{m_1+m_2}$	$\frac{I_{2y}-I_{2z}}{I_{1y}+I_{2y}}$	0	c
d_4	1	1	1	$\frac{2mr}{I_z}$	c

sensitive element, since its effect is either negligibly small or removed by the demodulation of the secondary oscillations.

Let us now finally look at the most simplified form of the generalised equations when the external angular rate Ω is small comparing to natural frequency and slowly varying, e.g. $\Omega^2 \approx 0$, $\Omega \approx$ const, and $\dot{\Omega} \approx 0$. As a result, we obtain the most commonly known form of CVG motion equations

$$\begin{cases} \ddot{x}_1 + 2\zeta_1\omega_1\dot{x}_1 + \omega_1^2 x_1 = q_1 - g_1\Omega\dot{x}_2, \\ \ddot{x}_2 + 2\zeta_2\omega_2\dot{x}_2 + \omega_2^2 x_2 = q_2 + g_2\Omega\dot{x}_1, \end{cases} \qquad (2.30)$$

where only Coriolis terms are present. Although system (2.30) is good enough to describe CVG principles of operations, we shall use Eq. (2.29) to analyse sensitive element motion, and to design signal processing and control algorithms. Needless to say that if any results obtained for the Eq. (2.29), they are applicable to CVG of any discussed above design.

Resume

Derived in this chapter, the generalised motion equations of Coriolis vibratory gyroscopes allow us to analyse motion and optimise the design of sensitive elements and the results can be equally applicable to different designs. For apparent reasons, not all possible designs were included in the Table 2.1. However, as demonstrated above, motion equations for any sensitive element design could be derived using Lagrange method, and then its essential terms can be related to the corresponding terms in the generalised form (2.29). As soon as proper entries to the Table 2.1 are identified, all the results and methodologies can be applied to the specific CVG design.

Chapter 3
Sensitive Element Dynamics

After CVG sensitive element motion equations were obtained in the previous chapter, the next step would be to find solutions of these equations. As it has been already mentioned, obtained equations are not quite suitable for analysis in a closed form in terms of the primary and secondary coordinates due to the presence of the unknown and variable angular rate as a coefficient. However, given certain assumptions and using the fact that useful information in vibratory gyroscopes is present in amplitudes and phases of oscillations, solutions suitable for analysis still can be obtained.

Analysis of the sensitive element dynamics on a fixed and rotating base will be presented in this chapter. Solutions in terms of amplitudes and phases of the primary and secondary oscillations will be obtained. Sensitive element motion trajectory analysis will be presented as well. Finally, methodology of realistic numerical simulation of CVG dynamics will be explained in detail.

3.1 Primary Motion of the Sensitive Element

Sometimes it is more convenient to view at the angular rate sensing by means of a Coriolis vibratory gyroscope as an amplitude modulation. Indeed, output secondary oscillations are the primary oscillations modulated by the external angular rate. This is apparent from the analysis of the generalised motion Eq. (2.29). First equation describes primary oscillations that are induced by the excitations system and is coupled with the second motion equation via Coriolis term, which is linearly related to the external angular rate.

If there is no external rotation ($\Omega = 0$) the motion equations become independent set of two unconnected equations

$$\begin{bmatrix} \ddot{x}_1 + 2\zeta_1 \omega_1 \dot{x}_1 + \omega_1^2 x_1 = q_{10} \sin(\omega t), \\ \ddot{x}_2 + 2\zeta_2 \omega_2 \dot{x}_2 + \omega_2^2 x_2 = 0. \end{bmatrix} \tag{3.1}$$

Here there are no external forces are applied to the secondary motion, and q_{10} is the amplitude of the accelerations from the excitations system, ω is the excitation

© Springer International Publishing Switzerland 2016
V. Apostolyuk, *Coriolis Vibratory Gyroscopes*,
DOI 10.1007/978-3-319-22198-4_3

frequency. From Eq. (3.1) it is apparent that without external rotation secondary motion is absent, and only the first equation needs to be solved.

Closed form solutions for Eq. (3.1) are

$$x_1(t) = C_1 e^{-\zeta_1 t} \sin\left(t\omega_1\sqrt{1-\zeta_1^2} + \varphi_1\right)$$
$$+ \frac{q_{10}}{\sqrt{\left(\omega_1^2 - \omega^2\right)^2 + 4\omega_1^2\zeta_1^2\omega^2}}\sin(\omega t + \gamma), \qquad (3.2)$$
$$x_2(t) = 0.$$

Phase shift γ of the resulting primary oscillations is given by

$$tg(\gamma) = -\frac{2\zeta_1\omega_1\omega}{\omega_1^2 - \omega^2},$$

and constants C_1 and φ_1 are determined from the initial conditions.

Analysis of the obtained solutions (3.2) shows us that settled oscillations of the CVG sensitive element occur with an amplitude that is proportional to the excitation force after natural oscillations disappear due to the damping (see Fig. 3.1).

At the same time proof mass remains motionless, and output signal from the sensor will be zero.

Furthermore transient time of the natural oscillations will determine start-up time of CVG, when only settled oscillations of the sensitive element remain and occur with the constant amplitude.

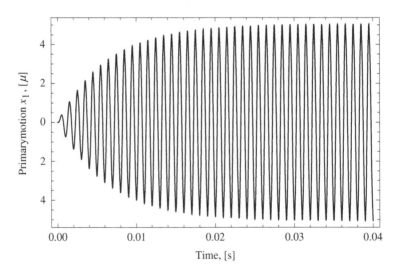

Fig. 3.1 Primary oscillations ($\omega_1 = 1000$ Hz, $\omega_1 = \zeta_2 = 0.025$, $\Omega = 0$)

$$A_{10} = \frac{q_{10}}{\sqrt{\left(\omega_1^2 - \omega^2\right)^2 + 4\omega_1^2\zeta_1^2\omega^2}}. \tag{3.3}$$

Since primary oscillations are the carrier that is modulated by the external angular rate, in order to perform reliable measurements, the carrier must be highly stable in terms of amplitude and frequency and its amplitude must be as high as possible.

The natural way to achieve the highest amplitude of the primary oscillations is to excite them in resonance with its eigenfrequency

$$\omega = \omega_1\sqrt{1 - 2\zeta^2}. \tag{3.4}$$

In case of resonance primary amplitude (3.3) becomes

$$A_{10} = \frac{q_{10}}{2\zeta_1\omega_1^2\sqrt{1 - \zeta_1^2}}. \tag{3.5}$$

Analysis of the expression (3.5) shows that the primary amplitude is higher when damping ζ_1 is lower, as well as the natural frequency ω_1 of the primary oscillations.

3.2 Sensitive Element Motion on a Rotating Base

Having looked at the primary oscillations of the sensitive element and methods of their efficient excitation, let us move on to the secondary oscillations. Studying solutions (3.2), one could see that if external angular rate is absent then secondary oscillations are absent as well.

Let us study now behaviour of the CVG sensitive element on the base that rotates with constant angular rate, e.g. $\dot{\Omega} = 0$. Motion Eq. (2.29) takes slightly simpler form

$$\begin{cases} \ddot{x}_1 + 2\zeta_1\omega_1\dot{x}_1 + (\omega_1^2 - d_1\Omega^2)x_1 = q_1 - g_1\Omega\dot{x}_2, \\ \ddot{x}_2 + 2\zeta_2\omega_2\dot{x}_2 + (\omega_2^2 - d_2\Omega^2)x_2 = q_2 + g_2\Omega\dot{x}_1. \end{cases} \tag{3.6}$$

Unlike the set of Eq. (3.1) here we have a system of equations that are cross-coupled by gyroscopic terms with the angular rate. Most importantly it becomes a system of ordinary differential equations with constant coefficients.

If no external forces affect secondary motion ($q_2 = 0$) and assuming ideal elastic suspension with no coupling, secondary motion will depend only on the angular rate Ω.

We are also interested in the forced solution of the system (3.6) since it is responsible for the angular rate measurement, while natural solution describes transient processes.

If excitation of the primary oscillation is harmonic, acceleration from the excitation forces could be represented in a complex form as

$$q_1(t) = \text{Re}\{q_{10}e^{i\omega t}\}. \tag{3.7}$$

Here ω is the excitation frequency and phase is assumed zero. Primary and secondary oscillations of the proof mass and the decoupling frame we shall search as a particular solution of the system (3.6) in the following form

$$
\begin{aligned}
x_1(t) &= \text{Re}\{A_1 e^{i\omega t}\}, \\
x_2(t) &= \text{Re}\{A_2 e^{i\omega t}\}.
\end{aligned}
\tag{3.8}
$$

We search here for the forced oscillations occurring at the excitation frequency ω, and parameters of these oscillations are described by the constant complex primary and secondary amplitudes A_1 and A_2 as

$$
\begin{aligned}
A_1 &= A_{10}e^{i\varphi_{10}}, \\
A_2 &= A_{20}e^{i\varphi_{20}}.
\end{aligned}
\tag{3.9}
$$

Here A_{10}, A_{20}, φ_{10}, and φ_{20} are the *constant* real amplitudes and phases of the primary and secondary oscillations, respectively. Substituting suggested solutions (3.8) into Eq. (3.6) we obtain the following system of algebraic equations in terms of complex amplitudes instead of differential equations:

$$
\begin{cases}
(\omega_1^2 - d_1\Omega^2 - \omega^2 + 2\zeta_1\omega_1 i\omega)A_1 + g_1 i\omega\Omega A_2 = q_{10}, \\
(\omega_2^2 - d_2\Omega^2 - \omega^2 + 2\zeta_2\omega_2 i\omega)A_2 - g_2 i\omega\Omega A_1 = 0.
\end{cases}
\tag{3.10}
$$

Solutions of the system (3.10) with respect to complex amplitudes are

$$
\begin{aligned}
A_1 &= \frac{q_{10}(\omega_2^2 - d_2\Omega^2 - \omega^2 + 2\zeta_2\omega_2 i\omega)}{\Delta}, \\
A_2 &= \frac{g_2 q_{10} i\omega}{\Delta}\Omega, \\
\Delta &= (\omega_1^2 - d_1\Omega^2 - \omega^2)(\omega_2^2 - d_2\Omega^2 - \omega^2) \\
&\quad - (g_1 g_2\omega^2\Omega^2 + 4\zeta_1\zeta_2\omega_1\omega_2\omega^2) \\
&\quad + 2i\omega\big[\zeta_1\omega_1(\omega_2^2 - d_2\Omega^2 - \omega^2) + \zeta_2\omega_2(\omega_1^2 - d_1\Omega^2 - \omega^2)\big]
\end{aligned}
\tag{3.11}
$$

Using of the complex amplitude method, which is modification of the method of averaging, results in possibility to analyse amplitudes and phases of the sensitive element motion instead of analysing in terms of its displacements x_1 and x_2. The former are of great interest from the angular rate sensing point of view.

Looking at (3.11) one can see that the amplitude of the secondary oscillations is almost linear function of the unknown angular rate. However, amplitudes (3.11) are the complex valued quantities. Conversion from the complex amplitudes to the real amplitudes and phases is performed as follows:

$$A_{i0} = |A_i| = \sqrt{\mathrm{Re}^2 A_i + \mathrm{Im}^2 A_i},$$
$$\tan \varphi_{i0} = \frac{\mathrm{Im}\, A_i}{\mathrm{Re}\, A_i}. \tag{3.12}$$

Here subscript $i = 1, 2$ for the primary and secondary amplitude and phase. Applying transformations (3.12) to the complex amplitudes (3.11) results in

$$A_{10} = \frac{q_{10}\sqrt{(\omega_2^2 - d_2\Omega^2 - \omega^2)^2 + 4\zeta_2^2\omega_2^2\omega^2}}{|\Delta|},$$
$$A_{20} = \frac{g_2 q_{10}\omega}{|\Delta|}\Omega, \tag{3.13}$$

where

$$|\Delta|^2 = \left[(\omega_1^2 - d_1\Omega^2 - \omega^2)(\omega_2^2 - d_2\Omega^2 - \omega^2)\right.$$
$$\left. - \omega^2(g_1 g_2\Omega^2 + 4\zeta_1\zeta_2\omega_1\omega_2)\right]^2$$
$$+ 4\omega^2\left[\zeta_1\omega_1(\omega_2^2 - d_2\Omega^2 - \omega^2) + \zeta_2\omega_2(\omega_1^2 - d_1\Omega^2 - \omega^2)\right]^2.$$

The real phases of the primary and secondary oscillations are given by the following expressions:

$$\mathrm{tg}(\phi_1) = \frac{2\omega\left[(\omega_2^2 - d_2\Omega^2 - \omega^2)b_1 + \omega_2\zeta_2 b_2\right]}{(\omega_2^2 - d_2\Omega^2 - \omega^2)b_2 - 4\omega_2\zeta_2\omega^2 b_1},$$
$$\mathrm{tg}(\phi_2) = \frac{(\omega_1^2 - d_1\Omega^2 - \omega^2)(\omega_2^2 - d_2\Omega^2 - \omega^2) - \omega^2(4\zeta_1\zeta_2\omega_1\omega_2 + g_1 g_2\Omega^2)}{2\omega\left[\omega_1\zeta_1(\omega_2^2 - d_2\Omega^2 - \omega^2) + \omega_2\zeta_2(\omega_1^2 - d_1\Omega^2 - \omega^2)\right]},$$
$$\tag{3.14}$$

where

$$b_1 = \omega_1\zeta_1(\omega_2^2 - d_2\Omega^2 - \omega^2) + \omega_2\zeta_2(\omega_1^2 - d_1\Omega^2 - \omega^2),$$
$$b_2 = (\omega_1^2 - d_1\Omega^2 - \omega^2)(\omega_2^2 - d_2\Omega^2 - \omega^2) - \omega^2(4\zeta_1\zeta_2\omega_1\omega_2 + g_1 g_2\Omega^2).$$

Using formulae (3.13) and (3.14) we can now study how CVG sensitive element responds to the external angular rate and how it could be measured with the highest possible efficiency.

Amplitude of secondary oscillations A_{20} (or secondary amplitude), as a function of the excitation frequency ω, is shown in Fig. 3.2.

It is apparent that maximum response to the constant angular rate is achieved when resonance with the lower natural frequency occurs. At the same time, the lower is damping, the higher are peaks of the secondary amplitude.

As it becomes apparent from (3.13) and (3.14), amplitude-related performances of a CVG are strongly related to such parameters of its sensitive element as natural frequencies and damping factors, as well as type related coefficients d_i. In order to make analysis of the CVG sensitive element motion more intuitive, let us introduce the following new variables: $k = \omega_1$ is the natural frequency of the primary motion, $\delta k = \omega_2/\omega_1$ is the relative natural frequency of the secondary motion, $\delta\omega = \omega/k$ is the relative excitation frequency, $\delta\Omega = \Omega/k$ is the relative angular rate, which is apparently small in comparison to primary natural frequency ($\Omega \ll k$, $\delta\Omega \ll 1$). In terms of these new variables, expressions (3.13) for the primary and secondary amplitudes can be rewritten as

$$
A_{10} = \frac{q_{10}k^2\sqrt{(\delta k^2 - d_2\delta\Omega^2 - \delta\omega^2)^2 + 4\zeta_2^2\delta k^2\delta\omega^2}}{|\Delta|},
$$

$$
A_{20} = \frac{g_2 q_{10}\delta\omega}{|\Delta|}\delta\Omega,
$$

(3.15)

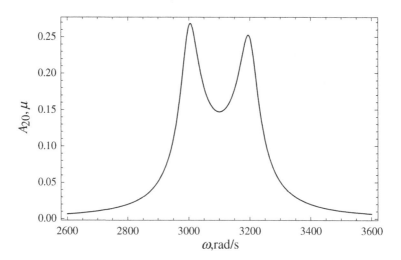

Fig. 3.2 Secondary amplitude as a function of the excitation frequency ($\omega_1 = 3000$, $\omega_2 = 3200$, $\zeta_1 = \zeta_2 = 0.01$, $\Omega = 1$)

and squared denominator is

$$
\begin{aligned}
|\Delta|^2 = k^8 \big[& (1 - d_1 \delta\Omega^2 - \delta\omega^2)(\delta k^2 - d_2 \delta\Omega^2 - \delta\omega^2) \\
& - \delta\omega^2 (g_1 g_2 \delta\Omega^2 + 4\zeta_1 \zeta_2 \delta k) \big]^2 \\
& + 4k^8 \delta\omega^2 \big[\zeta_1 (\delta k^2 - d_2 \delta\Omega^2 - \delta\omega^2) + \zeta_2 \delta k (1 - d_1 \delta\Omega^2 - \delta\omega^2) \big]^2.
\end{aligned}
$$

Expressions (3.15) can be simplified even further, if small relative angular rate is assumed, e.g. $\delta\Omega \ll 1$ and $\delta\Omega^2 \approx 0$:

$$
\begin{aligned}
A_{10} &\approx \frac{q_{10}}{k^2 \sqrt{(1 - \delta\omega^2)^2 + 4\zeta_1^2 \delta\omega^2}}, \\
A_{20} &\approx \frac{g_2 q_{10} \delta\omega}{k^2 \sqrt{(\delta k^2 - \delta\omega^2)^2 + 4\zeta_2^2 \delta k^2 \delta\omega^2} \sqrt{(1 - \delta\omega^2)^2 + 4\zeta_1^2 \delta\omega^2}} \delta\Omega \\
&\approx \frac{g_2 A_{10} \delta\omega}{\sqrt{(\delta k^2 - \delta\omega^2)^2 + 4\zeta_2^2 \delta k^2 \delta\omega^2}} \delta\Omega.
\end{aligned}
\tag{3.16}
$$

Expressions for the phases (3.14) are also can be rewritten in terms of relative sensitive element parameters. In case of small relative angular rate they become

$$
\begin{aligned}
\mathrm{tg}(\varphi_1) &= \frac{2\delta\omega\zeta_1}{1 - \delta\omega}, \\
\mathrm{tg}(\varphi_2) &= \frac{\delta k^2 - (1 + 4\zeta_1 \zeta_2 \delta k + \delta k^2)\delta\omega^2 + \delta\omega^4}{2\delta k \delta\omega(\zeta_2 + \zeta_1 \delta k) - 2\delta\omega^3(\zeta_1 + \zeta_2 \delta k)}.
\end{aligned}
\tag{3.17}
$$

Expressions (3.16) and (3.17) allow to analyse motion of the CVG sensitive element in the presence of external rotation using dimensionless parameters, which gives a certain level of generalisation of obtained results.

Another important aspect of motion equation analysis is the characteristic equation that gives eigenfrequencies of the sensitive element. In order to obtain characteristic equation, the differential operator $s = d/dt$ is used to transform the motion Eq. (3.6) to the following form:

$$
\begin{cases}
s^2 x_1 + 2\zeta_1 \omega_1 s x_1 + (\omega_1^2 - d_1 \Omega^2) x_1 = q_1 - g_1 \Omega s x_2, \\
s^2 x_2 + 2\zeta_2 \omega_2 s x_2 + (\omega_2^2 - d_2 \Omega^2) x_2 = q_2 + g_2 \Omega s x_1.
\end{cases}
\tag{3.18}
$$

This transformed equation can be then rewritten in a matrix form as

$$
A \cdot \begin{bmatrix} x_1 \\ x_2 \end{bmatrix} = \begin{bmatrix} q_1 \\ q_2 \end{bmatrix}.
\tag{3.19}
$$

where A is the principal system matrix given defined by the Eq. (3.18) terms as

$$
A = \begin{bmatrix} s^2 + 2\zeta_1\omega_1 s + \omega_1^2 - d_1\Omega^2 & g_1\Omega s \\ -g_2\Omega s & s^2 + 2\zeta_2\omega_2 s + \omega_2^2 - d_2\Omega^2 \end{bmatrix}. \tag{3.20}
$$

Characteristic equation is defined as

$$
\det A = 0. \tag{3.21}
$$

Substituting expressions (3.20) into Eq. (3.21) yields the characteristic equation as

$$
s^4 + a_3 s^3 + a_2 s^2 + a_1 s + a_0 = 0. \tag{3.22}
$$

where coefficients a_j are

$$
\begin{aligned}
a_3 &= 2s^3(\zeta_1\omega_1 + \zeta_2\omega_2), \\
a_2 &= s^2(\omega_1^2 - d_1\Omega^2 + \omega_2^2 - d_2\Omega^2 + 4\zeta_1\omega_1\zeta_2\omega_2 + g_1 g_2\Omega^2), \\
a_1 &= 2s[\zeta_1\omega_1(\omega_2^2 - d_2\Omega^2) + \zeta_2\omega_2(\omega_1^2 - d_1\Omega^2)], \\
a_0 &= (\omega_1^2 - d_1\Omega^2)(\omega_2^2 - d_2\Omega^2).
\end{aligned}
$$

Characteristic Eq. (3.22) allows analysis of stability of the sensitive element oscillations by using Routh criteria as

$$
\begin{aligned}
a_1 a_2 a_3 - a_1^2 - a_3^2 a_0 &> 0, \\
a_j &> 0.
\end{aligned} \tag{3.23}
$$

Coefficients of the characteristic Eq. (3.22) are functions of the angular rate, damping factors and natural frequencies of the system, and only angular rate is unknown. All other parameters are subjects to design in accordance to the stability conditions (3.23). Analysing coefficients of (3.22) one can find that stable secondary oscillations occur when the external angular rate is less than natural frequency of primary oscillations:

$$
-\omega_1 < \Omega < \omega_1. \tag{3.24}
$$

Relationships (3.24) usually are satisfied, since external angular rate is much less than the natural frequency of primary oscillations.

Closed form solution of the complete fourth-order Eq. (3.22) is quite complicated and therefore useless for subsequent analysis. However, if damping is assumed to be negligibly small ($\zeta_1 = \zeta_2 = 0$), then all odd order terms will disappear and characteristic Eq. (3.22) becomes bi-quadratic, which is being rewritten using dimensionless variable introduced earlier, is as follows:

$$s^4 + s^2 k^2 (1 - d_1 \delta \Omega^2 + \delta k^2 - d_2 \delta \Omega^2 + g_1 g_2 \delta \Omega^2)$$
$$+ k^4 (1 - d_1 \delta \Omega^2)(\delta k^2 - d_2 \delta \Omega^2) = 0. \tag{3.25}$$

Roots of Eq. (3.25) are relating to the relative eigenfrequencies $\delta \omega_{j0}$ of the primary and secondary oscillations as

$$s_{1,2} = \pm i k \delta \omega_{10},$$
$$s_{3,4} = \pm i k \delta \omega_{20}. \tag{3.26}$$

Solving Eq. (3.25) results in the following expressions for the eigenfrequencies:

$$\delta \Omega_{j0}^2 = \frac{1}{2} \left[1 + \delta k^2 - (d_1 + d_2 - g_1 g_2) \delta \Omega^2 \right]$$
$$+ \frac{(-1)^j}{2} \left\{ 4(\delta k^2 - d_1 \delta \Omega^2)(d_2 \delta \Omega^2 - 1) \right.$$
$$\left. + \left[1 + \delta k^2 - (d_1 + d_2 - g_1 g_2) \delta \Omega^2 \right]^2 \right\}^{1/2} \tag{3.27}$$

Graphic plot of the frequencies (3.27) as a functions of the external angular rate is shown in Fig. 3.3, where relative natural frequency is $\delta k = 1.05$.

When there is no external rotation, eigenfrequencies become equal to corresponding natural frequencies. When external rotation is applied, these frequencies start to shift due to the angular rate. Although this dependence does not appear to be linear, it becomes very close to linear when natural frequencies are equal ($\delta k = 1$).

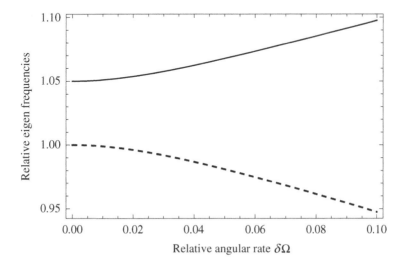

Fig. 3.3 Relative eigenfrequencies (*dashed* primary, *solid* secondary)

Such a frequency modulation could be considered as complementary to amplitude modulation, demonstrated by (3.15), and used to improve angular rate measurements.

3.3 Modelling Proof Mass Motion Trajectory

It is possible to view the CVG sensitive element as a two-dimensional pendulum. In presence of the constant external angular rate, trajectory of its centre of gravity forms an ellipse, as shown in Fig. 3.4.

In this figure, a and b are the big and small half-axes of the ellipse, θ is the angle of the ellipse rotation relatively to the axes of primary x_1 and secondary x_2 oscillations. It is well known that these parameters (namely half-axes and angle of rotation) depend on amplitudes and phases of primary and secondary oscillations, which in turn depend on parameters of the sensitive element design and unknown angular rate.

The problem, which is to be addressed in this section, is to develop and analyse a mathematical model of the ellipse parameters as a function of the sensitive element design and its parameters.

Without loss of generality, primary and secondary coordinates of the CVG sensitive element in its steady motion can be represented as

$$
\begin{aligned}
x_1(t) &= A_{10} \cos \omega t, \\
x_2(t) &= A_{20} \cos(\omega t + \varphi) = A_{20}[\cos \omega t \cos \varphi - \sin \omega t \sin \varphi],
\end{aligned}
\tag{3.28}
$$

Fig. 3.4 Sensitive element motion trajectory

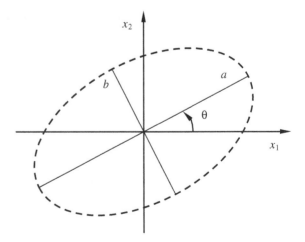

where A_{10} and A_{20} are the primary and secondary amplitudes, φ is the phase shift between the primary and secondary oscillations, ω is the oscillations circular frequency.

First of all, we have to exclude terms containing oscillation phase ωt from Eq. (3.28). In order to achieve this, we express sine and cosine of the ωt from the first equation and substitute them into the second one, yielding

$$\frac{x_1}{A_{10}}\cos\varphi - \frac{x_2}{A_{20}} = \pm\sqrt{1 - \frac{x_1^2}{A_{10}^2}\sin\varphi}. \qquad (3.29)$$

Squaring both sides of the Eq. (3.29) results in

$$\frac{x_1^2}{A_{10}^2} + \frac{x_2^2}{A_{20}^2} - \frac{2x_1 x_2 \cos\varphi}{A_{10}A_{20}} = \sin\varphi^2. \qquad (3.30)$$

Assuming X_1 and X_2 as the coordinates of the sensitive element centre of gravity in the coordinate system that is rotated by the angle θ, it is easy relating them to the original coordinates as

$$\begin{aligned} x_1 &= X_1 \cos\theta - X_2 \sin\theta, \\ x_2 &= X_1 \sin\theta + X_2 \cos\theta. \end{aligned} \qquad (3.31)$$

In terms of these coordinates ellipse equation has its conventional form

$$\frac{X_1^2}{a^2} + \frac{X_2^2}{b^2} = 1.$$

Substituting (3.31) into (3.30) results in

$$\begin{aligned} X_1^2 &\frac{A_{20}^2 \cos^2\theta - A_{10}A_{20}\cos\varphi \sin 2\theta + A_{10}^2 \sin^2\theta}{A_{10}^2 A_{20}^2} \\ &+ X_1 X_2 \frac{(A_{10}^2 - A_{20}^2)\sin 2\theta - 2A_{10}A_{20}\cos\varphi \cos 2\theta}{A_{10}^2 A_{20}^2} \\ &+ X_2^2 \frac{A_{10}^2 \cos^2\theta + A_{10}A_{20}\cos\varphi \sin 2\theta + A_{20}^2 \sin^2\theta}{A_{10}^2 A_{20}^2} = \sin^2\varphi. \end{aligned} \qquad (3.32)$$

Apparently, in order to transform Eq. (3.32) to the conventional form, the second term must disappear. This will occur if θ satisfies the following condition

$$\theta = \frac{1}{2}\tan^{-1}\frac{2A_{10}A_{20}\cos\varphi}{A_{10}^2 - A_{20}^2}. \qquad (3.33)$$

If angle θ is defined be (3.33), then half-axes of the ellipse will be given by the following expressions:

$$a = \frac{A_{10}A_{20}\sin\varphi}{\sqrt{A_{20}^2\cos^2\theta - A_{10}A_{20}\cos\varphi\sin 2\theta + A_{10}^2\sin^2\theta}},$$

$$b = \frac{A_{10}A_{20}\sin\varphi}{\sqrt{A_{10}^2\cos^2\theta + A_{10}A_{20}\cos\varphi\sin 2\theta + A_{20}^2\sin^2\theta}}. \tag{3.34}$$

Now we have to express half-axes a and b of the ellipse, and its angle of rotation θ in terms of the sensitive element parameters and external angular rate.

In perfectly tuned CVG, where primary and secondary eigenfrequencies are perfectly matched, phase shift is zero ($\varphi = 0$). This allows us to approximate expressions (3.33) and (3.34) with the linear terms of its Taylor series representation around zero phase shift:

$$a \approx \frac{A_{10}A_{20}\varphi}{A_{20}\cos\theta - A_{10}\sin\theta},$$

$$b \approx \frac{A_{10}A_{20}\varphi}{A_{10}\cos\theta + A_{20}\sin\theta}. \tag{3.35}$$

One should note that in approximations (3.35) zero phase shift results in zero big half-axis a, which apparently is not acceptable in trajectory analysis applications. Let us have a close look at the dependencies (3.34) for the elliptical trajectory of the CVG sensitive element. Obviously, expression for the big half-axis has a peculiarity in vicinity of the zero phase shift, when the denominator of the formula may become zero. However, far from the zero phase, this function is quite smooth and has no such peculiarities. From the extensive analysis of this dependency the following simple approximation is suggested:

$$a \approx A_{10} + \frac{A_{20}^2}{4}(1 + \cos 2\varphi). \tag{3.36}$$

Dimensionless relative error of the approximation (3.36) is shown in Fig. 3.5.

Observable singularities along the zero phase shift in Fig. 3.5 demonstrate that approximation is more relevant than the initial formula. It is easy to see that approximation (3.36) is certainly accurate enough for small phase shifts and small relative secondary amplitudes, which are typical for the CVG sensitive element motions.

Let us now analyse how elliptical trajectory parameters depend on the external angular rate and sensitive element characteristics.

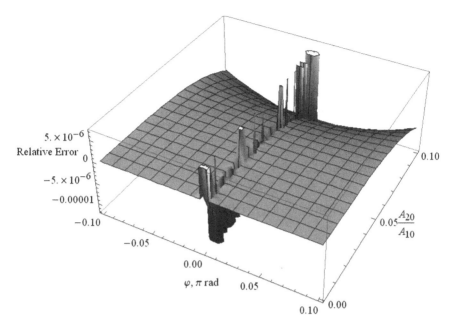

Fig. 3.5 Relative error of the big half-axis approximation

One should note that the sensitive element trajectory parameters depend on the phase shift $\varphi = \varphi_2 - \varphi_1$ between primary and secondary phases. Most importantly, based upon (3.17), phases do not depend on the angular rate.

In case of primary resonance ($\delta\omega = 1$), sine and cosine of this phase shift can be calculated as

$$
\begin{aligned}
\sin\varphi &= \frac{1 - \delta k^2}{R}, \\
\cos\varphi &= \frac{2\zeta_2 \delta k}{R}, \\
R &= \sqrt{\delta k^4 - 2(1 - 2\zeta_2^2)\delta k^2 + 1}.
\end{aligned}
\tag{3.37}
$$

Expressions (3.16) and (3.17) along with the phase shift representations (3.37) can now be used to analyse parameters of the actual trajectory of the CVG sensitive element.

Let us now substitute expressions (3.16) and (3.37) into the formula for the big half-axis approximation (3.36):

$$
a = \frac{q_{10}(4k^2 R^2 \zeta_1 + g_2^2 q_{10} \delta\Omega^2)}{8k^4 R^2 \zeta_1^2}.
$$

Remembering that the relative angular rate is small ($\delta\Omega^2 \approx 0$), big half-axis, as expected, becomes the primary amplitude:

$$a \approx \frac{q_{10}}{2k^2\zeta_1} = A_{10}|_{\delta\Omega=1}. \qquad (3.38)$$

Small half-axis of the sensitive element trajectory after substitution is

$$b = \frac{q_{10}g_2(1 - \delta k^2)}{\sqrt{2}k^2R\zeta_1\sqrt{R^2 + g_2^2\delta\Omega^2 + \sqrt{R^4 - 2g_2^2(R^2 - 8\zeta_2^2\delta k^2)\delta\Omega^2 + g_2^4\delta\Omega^4}}}\delta\Omega,$$

and dropping higher powers of the angular rate results in

$$b = \frac{q_{10}g_2(1 - \delta k^2)}{2k^2\zeta_1[1 - 2(1 - 2\zeta_2^2)\delta k^2 + \delta k^4]}\delta\Omega. \qquad (3.39)$$

Analysing formula (3.39) is apparent that small half-axis is absent in either of two cases: perfect match of the natural frequencies ($\delta k = 1$) or absence of the external angular rate ($\delta\Omega = 0$).

Finally, the last but not the least, angle of the trajectory rotation θ can be calculated using the following expression:

$$\theta = \frac{1}{2}\arctan\left[\frac{4g_2\zeta_2\delta k\delta\Omega}{1 - 2(1 - 2\zeta_2^2)\delta k^2 + \delta k^4}\right]. \qquad (3.40)$$

Similarly to the previous case, we can linearly approximate formula (3.40) for small angular rates:

$$\theta \approx \frac{2g_2\zeta_2\delta k}{1 - 2(1 - 2\zeta_2^2)\delta k^2 + \delta k^4}\delta\Omega. \qquad (3.41)$$

Or, in case of perfectly matched primary and secondary natural frequencies ($\delta k = 1$), (3.41) becomes

$$\theta \approx \frac{g_2}{2\zeta_2}\delta\Omega. \qquad (3.42)$$

Both accurate expression (3.40) and its linear approximation (3.42) are shown in Fig. 3.6.

From the graphs in Fig. 3.6 one can see that trajectory rotation angle is almost linear function of the unknown angular rate, which allows to use it to measure angular rate.

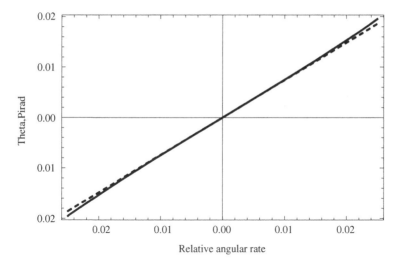

Fig. 3.6 Trajectory rotation angle as a function of angular rate (*solid* accurate, *dashed* linear approximation)

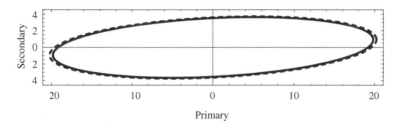

Fig. 3.7 CVG Trajectory simulation (*solid* simulated, *dashed* theoretical)

Figure 3.7 demonstrates how trajectory approximations (3.38), (3.39) and (3.42) related to the realistically simulated trajectory, based on the numerical solution of the original motion equations.

Here the solid line demonstrates steady motion trajectory of the simulated CVG sensitive element, and the dashed line corresponds to the trajectory, generated from the obtained trajectory parameters.

Comparison with the results of numerical simulation demonstrates high accuracy of the obtained mathematical model of the sensitive element motion trajectory. These dependencies allow not only further analysis of the sensitive element motion, but efficient synthesis of many different control loops improving overall CVG performance as well.

3.4 Numerical Simulation of CVG Dynamics Using Simulink®

Accurate numerical simulation of CVG is essential for proper verification of the developed mathematical models. Generalised Simulink model for CVG simulation in an open-loop operation mode is presented in Fig. 3.8.

Here block "Angular Rate" provides angular rate as an input to the system, block "Excitation" provides sinusoidal signal to the primary mode input, "Process Noise" can be added to the secondary mode input, and the "Sensor Noise" can be added to the CVG output. Secondary coordinate x_2 must be demodulated by the "Secondary Detector" to remove primary carrier signal.

To make the simulation results as realistic as possible, the following most generalised sensitive element motion Eq. (2.29) simplified for small angular rates ($\Omega^2 \approx 0$) will be used:

$$\begin{cases} \ddot{x}_1 + 2\zeta_1\omega_1\dot{x}_1 + \omega_1^2 x_1 = q_1 - g_1\Omega\dot{x}_2 - d_3\dot{\Omega}x_2, \\ \ddot{x}_2 + 2\zeta_2\omega_2\dot{x}_2 + \omega_2^2 x_2 = q_2 + g_2\Omega\dot{x}_1 + d_4\dot{\Omega}x_1. \end{cases} \tag{3.43}$$

Corresponding simulation model (contents of the "Sensitive Element Dynamics" subsystem block in Fig. 3.8) is shown in Fig. 3.9.

Primary and secondary dynamics in this model is simulated using transfer function block "Transfer Fcn" from Simulink. Parameters of the sensitive element are replaced with the following variables: k1 = ω_1, k2 = ω_2, h1 = ζ_1, h2 = ζ_2, etc.

Demodulation of the secondary output is performed by the synchronous demodulator ("Secondary Detector" block in Fig. 3.8) as shown in Fig. 3.10.

Secondary output is multiplied by the sinusoidal signal at the excitation frequency and the result is passed through the low-pass eighth-order Butterworth filter. Filter output is multiplied with the factor, which scales it to the rate-like output.

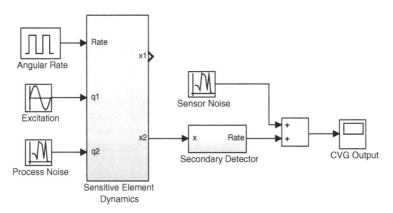

Fig. 3.8 CVG simulation model

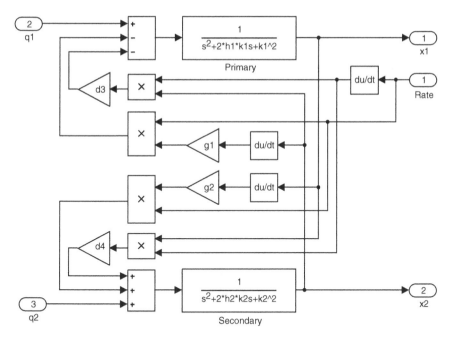

Fig. 3.9 Sensitive element dynamics model

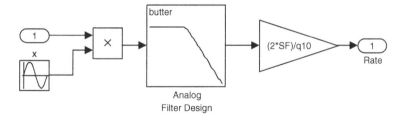

Fig. 3.10 Secondary detector model

Resume

As demonstrated in this chapter, solutions of the CVG motion equations not only give useful insights into how the sensitive element reacts to the external rotation, but also allow us to calculate and optimise the main performances of these types of gyroscopes.

Attentive readers may notice that we did not actually solve the original motion equations, but obtained steady state solutions for amplitudes, phase, and eigenfrequencies related to the external angular rate. This was done due to the fact that closed formed solutions of the original equations in terms of primary and secondary motions, which are oscillatory, are complicated and therefore useless for further analysis. The good news is we do not need them, since all the important dependencies from angular rate express themselves in amplitudes and phases, rather than actual sensitive element motions. However, steady state solutions, related to constant angular rate, do not allow analysis of the sensitive element behaviour when angular rate varies in time. And this problem is yet to be solved in the next chapter.

Chapter 4
CVG Dynamics in Demodulated Signals

In the previous chapter we analysed dynamics of the CVG sensitive element in terms of generalised coordinates x_1 and x_2 that describe harmonic primary and secondary motion correspondingly.

Being based on sensing of Coriolis acceleration due to the rotation in oscillating structures, CVGs have a lot more complicated mathematical models, compared to the conventional types of gyroscopes. One such complication is a result of the useful signal proportional to the external angular rate being modulated with the intentionally excited primary oscillations. From the mathematical modelling point of view, this leads to necessity to "demodulate" the solution in terms of the sensitive element displacements to obtain practically feasible insights into CVG dynamics and errors. From the control systems point of view, conventional representation of CVGs incorporates primary oscillation excitation signal as an input to the dynamic system, and unknown angular rate as a coefficient of its transfer functions. As a result, dynamics of CVGs has been analysed mainly in steady state, while transient process analysis has been omitted due to its apparent complexity.

4.1 Motion Equations in Demodulated Signals

Simplified with respect to small angular rates ($\Omega^2 \approx 0$) generalised motion equations of CVG (2.29) take the following form:

$$\begin{cases} \ddot{x}_1 + 2\zeta_1\omega_1\dot{x}_1 + \omega_1^2 x_1 = q_1 - g_1\Omega\dot{x}_2 - d_3\dot{\Omega}x_2, \\ \ddot{x}_2 + 2\zeta_2\omega_2\dot{x}_2 + \omega_2^2 x_2 = q_2 + g_2\Omega\dot{x}_1 + d_4\dot{\Omega}x_1. \end{cases} \tag{4.1}$$

In the motion Eq. (4.1), angular rate Ω is included as an unknown and variable coefficient rather than an input to the double oscillator system. Conventional control systems representation of such a dynamic system is shown in Fig. 4.1.

In order to identify the angular rate one must detect secondary oscillations of the sensitive element and measure its amplitude, which is approximately directly proportional to the angular rate, and phase, which gives the sign.

© Springer International Publishing Switzerland 2016
V. Apostolyuk, *Coriolis Vibratory Gyroscopes*,
DOI 10.1007/978-3-319-22198-4_4

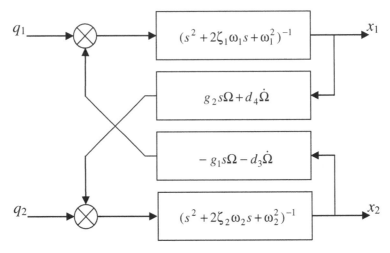

Fig. 4.1 Conventional representation of CVG in control systems

Compatible with the most control problems, CVG dynamics representation should have the unknown angular rate as an input and its measured value as an output.

Motion Eq. (4.1) can be further simplified by assuming that rotational and Coriolis accelerations, acting along the primary oscillation coordinate, are negligible in comparison to the accelerations from driving forces

$$g_1 \Omega \dot{x}_2 + d_3 \dot{\Omega} x_2 \ll q_1(t) \tag{4.2}$$

This assumption corresponds to cutting the feedback from secondary oscillations towards primary in Fig. 4.1. Assumption (4.2) results in the following simplified CVG motion equations:

$$\begin{cases} \ddot{x}_1 + 2\zeta_1 \omega_1 \dot{x}_1 + \omega_1^2 x_1 = q_1, \\ \ddot{x}_2 + 2\zeta_2 \omega_2 \dot{x}_2 + \omega_2^2 x_2 = g_2 \Omega \dot{x}_1 + d_4 \dot{\Omega} x_1. \end{cases} \tag{4.3}$$

Here we also assumed that no external driving forces are affecting the secondary oscillations, which means that $q_2(t) = 0$. System of Eq. (4.3) is now perfectly suitable for further transformations towards the desired representation in terms of the unknown angular rate.

By means of a proper chosen phase shift of the excitation voltage applied to the sensitive element, the excitation force could be shaped to the perfect harmonic form. Using exponential representation of complex numbers, such a driving force $q_1(t)$ could be represented as

$$q_1(t) = q_{10}\sin(\omega t) = \mathrm{Im}\{q_{10}e^{j\omega t}\} \qquad (4.4)$$

Here ω is the excitation frequency given in radians per second, q_{10} is the constant excitation acceleration amplitude.

Non-homogeneous solutions of the motion Eq. (4.3) for primary and secondary oscillations are searched similar to (3.8) and (3.9) to form

$$
\begin{aligned}
x_1(t) &= \mathrm{Im}\{A_1(t)e^{j\omega t}\}, \\
A_1(t) &= A_{10}(t)e^{j\varphi_1(t)}, \\
x_2(t) &= \mathrm{Im}\{A_2(t)e^{j\omega t}\}, \\
A_2(t) &= A_{20}(t)e^{j\varphi_2(t)},
\end{aligned}
\qquad (4.5)
$$

where A_{10} and A_{20} are the primary and secondary oscillation amplitudes, φ_{10} and φ_{20} are the corresponding phase shifts relative to the excitation force. Although these quantities are real (non-imaginary), they are combined in complex amplitude–phase variables A_1 and A_2. Note that contrary to the (3.8) and (3.9) these amplitudes and phases now vary in time.

Substituting expressions (4.4) and (4.5) into Eq. (4.3) results in the following motions equations in terms of the complex amplitude–phase variables rather than real generalised coordinates:

$$
\left\{
\begin{aligned}
&\ddot{A}_1 + 2(\zeta_1\omega_1 + j\omega)\dot{A}_1 + (\omega_1^2 - \omega^2 + 2j\omega\omega_1\zeta_1)A_1 = q_{10}, \\
&\ddot{A}_2 + 2(\zeta_2\omega_2 + j\omega)\dot{A}_2 + (\omega_2^2 - \omega^2 + 2j\omega\omega_2\zeta_2)A_2 \\
&\quad = (j\omega g_2\Omega + d_4\dot{\Omega})A_1 + g_2\dot{A}_1\Omega.
\end{aligned}
\right.
\qquad (4.6)
$$

Equation (4.6) describes variations of the amplitude and phase of the primary and secondary equations in time with respect to the unknown non-constant angular rate $\Omega(t)$. This allows conducting analysis of the Coriolis vibratory gyroscope dynamics without constraining the angular rate to be constant or slowly varying.

Analysing system (4.6), one can see that the first equation can be solved separately from the second one. After homogeneous solutions of the first equation faded out, only non-homogenous solution remains. In this case, steady amplitude of the primary oscillations is

$$A_1 = \frac{q_{10}}{\omega_1^2 - \omega^2 + 2j\omega_1\zeta_1\omega}, \qquad (4.7)$$

and it is constant in time, yielding $\ddot{A}_1 = \dot{A}_1 = 0$. Indeed, most of the time measurements of the angular rate are performed when primary oscillations have already settled. As a result, only equation for the secondary oscillations remains, in which the complex primary amplitude A_1 is just a constant parameter given by (4.7):

$$\ddot{A}_2 + 2(\zeta_2\omega_2 + j\omega)\dot{A}_2 + (\omega_2^2 - \omega^2 + 2j\omega\omega_2\zeta_2)A_2 = (j\omega g_2\Omega + d_4\dot{\Omega})A_1. \qquad (4.8)$$

Equation (4.8) now describes amplitude and phase of the secondary oscillations with respect to the settled primary oscillations.

4.2 CVG Transfer Functions

Having CVG sensitive element motion equation in the form (4.8), allows analysis of its transient processes in amplitudes and phases with respect to arbitrary angular rates applied to the system. Application of the Laplace transformation to Eq. (4.8) with respect to zero initial conditions for all time-dependent variables results in the following expressions:

$$[(s + j\omega)^2 + 2\zeta_2\omega_2(s + j\omega) + \omega_2^2]A_2(s) = A_1[d_4s + jg_2\omega]\Omega(s). \qquad (4.9)$$

Solution of the algebraic Eq. (4.9) for the secondary amplitude–phase Laplace transform is

$$A_2(s) = \frac{A_1(d_4s + jg_2\omega)}{(s + j\omega)^2 + 2\zeta_2\omega_2(s + j\omega) + \omega_2^2}\Omega(s). \qquad (4.10)$$

Considering the angular rate as an input, the system transfer function for the secondary amplitude–phase is

$$W_2(s) = \frac{A_2(s)}{\Omega(s)} = \frac{A_1(d_4s + jg_2\omega)}{(s + j\omega)^2 + 2\zeta_2\omega_2(s + j\omega) + \omega_2^2}$$

$$= \frac{q_{10}(d_4s + jg_2\omega)}{[(s + j\omega)^2 + 2\zeta_2\omega_2(s + j\omega) + \omega_2^2][\omega_1^2 - \omega^2 + 2j\omega\omega_1\zeta_1]}.$$
$$(4.11)$$

One should note that transfer function (4.11) has complex coefficients, which results in the complex system outputs as well. Although it is somewhat unusual, it still enables us to analyse CVG dynamics and transient processes due to the angular rate in an open-loop dynamic system.

Transfer function (4.11) describes demodulated dynamics of CVG in case of arbitrary changing secondary amplitudes. However, if angular rate is slowly varying compared to the primary oscillations, which is true for most of CVG applications, secondary amplitude can be considered slowly varying as well. This assumption allows us to neglect higher order derivatives of the secondary amplitude in Eq. (4.8), e.g. $\ddot{A}_2 \approx 0$. Neglecting the second-order derivative yields

$$2(\zeta_2\omega_2 + j\omega)\dot{A}_2 + (\omega_2^2 - \omega^2 + 2j\omega\omega_2\zeta_2)A_2 = (j\omega g_2\Omega + d_4\dot{\Omega})A_1, \quad (4.12)$$

and the corresponding angular rate transfer function becomes

$$W_2(s) = \frac{q_{10}(d_4s + jg_2\omega)}{[2\zeta_2\omega_2s + \omega_2^2 - \omega^2 + j2\omega(\zeta_2\omega_2 + s)][\omega_1^2 - \omega^2 + 2j\omega\omega_1\zeta_1]}. \quad (4.13)$$

Complex transfer function (4.13) is simpler in comparison to the function (4.11) and could replace it in certain specific problems when slowly-varying angular rate analysis is required.

While simulating dynamics of CVG based on the complex amplitude–phase transfer functions (4.11) or (4.13) one could have problems dealing with complex coefficients of these transfer functions. One way to avoid this problem is to consider real and imaginary parts of complex amplitude as separate signals, which are then combined together to produce real amplitude and phase. In order to obtain transfer functions for such signals let us represent primary and secondary amplitudes as:

$$\begin{aligned} A_1 &= A_{1R} + jA_{1I}, \\ A_2 &= A_{2R} + jA_{2I}. \end{aligned} \quad (4.14)$$

Primary oscillations components can be easily found by means of substituting expressions (4.14) into formula (4.7) thus resulting

$$\begin{aligned} A_{1R} &= \frac{q_{10}(\omega_1^2 - \omega^2)}{(\omega_1^2 - \omega^2)^2 + 4\omega_1^2\zeta_1^2\omega^2}, \\ A_{1I} &= -\frac{2q_{10}j\omega\omega_1\zeta_1}{(\omega_1^2 - \omega^2)^2 + 4\omega_1^2\zeta_1^2\omega^2}. \end{aligned} \quad (4.15)$$

At the same time, substituting expressions (4.15) into the motion Eq. (4.8), and applying Laplace transformation with zero initial conditions gives

$$\begin{cases} (\omega_2^2 - \omega^2 + 2\omega_2\zeta_2s + s^2)A_{2R}(s) - 2\omega(\omega_2\zeta_2 + s)A_{2I}(s) \\ \quad = (A_{1R}d_4s - A_{1I}g_2\omega)\Omega(s), \\ (\omega_2^2 - \omega^2 + 2\omega_2\zeta_2s + s^2)A_{2I}(s) + 2\omega(\omega_2\zeta_2 + s)A_{2R}(s) \\ \quad = (A_{1I}d_4s + A_{1R}g_2\omega)\Omega(s). \end{cases} \quad (4.16)$$

Resolving algebraic system (4.16) with respect to the unknown real and imaginary parts of the secondary complex amplitude results in

$$A_{2R}(s) = \frac{A_{1R}M_{RR}(s) + A_{1I}M_{RI}(s)}{P(s)}\Omega(s),$$

$$A_{2I}(s) = \frac{A_{1R}M_{IR}(s) + A_{1I}M_{II}(s)}{P(s)}\Omega(s). \tag{4.17}$$

Here the numerator polynomials from the real and imaginary parts of primary amplitudes are given by the following expressions:

$$
\begin{aligned}
M_{RR}(s) &= s(\omega_2^2 + 2\omega_2\zeta_2 s + s^2) - \omega^2(d_4 s - 2g_2(s + \omega_2\zeta_2)),\\
M_{RI}(s) &= \omega[2d_4 s(s + \omega_2\zeta_2) - g_2(\omega_2^2 - \omega^2 + 2\omega_2\zeta_2 s + s^2)],\\
M_{II}(s) &= 2\omega^2 g_2(s + \omega_2\zeta_2) + d_4 s(\omega_2^2 - \omega^2 + 2\omega_2\zeta_2 s + s^2)],\\
M_{IR}(s) &= \omega[g_2(\omega_2^2 - \omega^2 + 2\omega_2\zeta_2 s + s^2) - 2d_4 s(s + \omega_2\zeta_2)],\\
P(s) &= 4(s + \omega_2\zeta_2)^2\omega^2 + (\omega_2^2 - \omega^2 + 2\omega_2\zeta_2 s + s^2)^2.
\end{aligned}
\tag{4.18}
$$

Obtained expressions (4.15), (4.17), and (4.18) allow analysis of CVG dynamics in control system without necessity to involve complex-valued signals.

4.3 Amplitude and Phase Responses

In order to calculate the amplitude response of the system using transfer function (4.11), Laplace variable s must be replaced with the Fourier variable $j\lambda$, where λ is the frequency of the angular rate oscillations:

$$W_2(j\lambda) = \frac{jq_{10}(d_4\lambda + g_2\omega)}{[\omega_2^2 - (\lambda + \omega)^2 + 2j\zeta_2\omega_2(\lambda + \omega)][\omega_1^2 - \omega^2 + 2j\omega\zeta_1\omega_1]}. \tag{4.19}$$

Absolute value of the complex function (4.19) is the amplitude response of the secondary oscillations amplitude to the harmonic angular rate, and the corresponding phase of the complex function is the phase response:

$$A(\lambda) = \frac{q_{10}(d_4\lambda + g_2\omega)}{\sqrt{[(\omega_2^2 - (\lambda + \omega)^2)^2 + 4\zeta_2^2\omega_2^2\lambda + \omega)^2][(\omega_1^2 - \omega^2)^2 + 4\zeta_1^2\omega_1^2\omega^2]}},$$

$$\varphi(\lambda) = \tan^{-1}\left\{\frac{[\omega_2^2 - (\lambda + \omega)^2][\omega_1^2 - \omega^2] - 4\omega_1\omega_2\zeta_1\zeta_2\omega(\lambda + \omega)}{2[\omega_2\zeta_2(\lambda + \omega)(\omega_1^2 - \omega^2) + \omega_1\zeta_1\omega(\omega_2^2 - (\lambda + \omega)^2)]}\right\}. \tag{4.20}$$

One should note that assuming constant angular rate ($\lambda = 0$) in the expressions (4.20) the derived earlier expressions for the amplitude and phase of the secondary oscillations could be obtained.

Analysis of the expressions (4.20) shows that effect from the oscillating angular rate is practically equivalent to shift of the excitation frequency by the frequency of the angular rate. This causes CVGs, especially those with high Q-factor, to loose its resonant tuning, which in turn results in significant variation of its scale factor (dynamic error). Solution of this problem by means of proper choice of natural frequency split and damping will be considered later in this book.

4.4 Stability and Transient Process Optimisation

Both stability and unit-step transient process quality depend on position of the system poles in the real–imaginary plane. CVG operation in demodulated signals is described by the derived earlier system transfer function (4.11).

Poles of the transfer function (4.11) are as follows:

$$s_{1,2} = -\omega_2 \zeta_2 \pm j\omega_2 \sqrt{1 - \zeta_2^2} - j\omega. \tag{4.21}$$

Analysing expressions (4.21), it is easy to see that CVGs are inherently stable. Indeed, if the relative damping coefficient $\zeta_2 \leq 1$, then real parts of the poles are

$$-\omega_2 \zeta_2 < 0$$

If the relative damping coefficient $\zeta_2 > 1$, then real parts are

$$-\omega_2 \left(\zeta_2 \pm \sqrt{\zeta_1^2 - 1} \right) < 0$$

Ideal (half-oscillatory) unit-step angular rate transient process in secondary oscillations amplitude is achievable if imaginary parts of the poles (4.21) are zero. One pole has large imaginary part

$$-\omega_2 \sqrt{1 - \zeta_2^2} - \omega < 0,$$

which is always way below zero, and corresponds to high frequency oscillations in the envelope. The second pole is responsible for the low frequency oscillations, and is the most essential for the transient process. For this pole the ideal transient process condition has the following form:

$$\omega_2 \sqrt{1 - \zeta_2^2} - \omega = 0 \Rightarrow \omega_2 = \frac{\omega}{\sqrt{1 - \zeta_2^2}}. \tag{4.22}$$

For example, if primary oscillations are excited in pure resonance for better sensitivity, Eq. (4.22) is transformed to

$$\omega_2 = \omega_1 \sqrt{\frac{1 - 2\zeta_1^2}{1 - \zeta_2^2}}. \tag{4.23}$$

As a result, in order to provide ideal transient process for the secondary oscillations amplitude, one should design sensitive element of CVG with the natural frequency of the secondary oscillations according to (4.23).

Another important performance feature of a system transient process is its settling time, which is defined by the real part of the system poles and can be approximated as

$$T = -\frac{\ln(\varepsilon)}{\omega_2 \zeta_2}. \tag{4.24}$$

Here ε is the error tolerance ($\varepsilon = 0.01$ for 1 % tolerance). From this dependence one can see, that in order to minimise settling time, denominator $\omega_2 \zeta_2$ must be maximised. Since sensitivity of CVG is inversely related to its natural, reducing its damping along with the natural frequencies will increase its transient process settling time.

Another consequence of the presented above analysis of the system poles and its transient process is that actual amplitude of the secondary oscillations is mainly defined by the low frequency pole, while effect from the high frequency pole can be neglected, since it will be removed during demodulation process. In other words, predominant behaviour is a slow variation of the amplitude and phase, which is represented by the system transfer function (4.13). Single pole of this transfer function is

$$\begin{aligned}
s_1 &= -\frac{\omega_2^2 - \omega^2 + 2j\omega\zeta_2\omega_2}{2(\zeta_2\omega_2 + j\omega)} \\
&= -\omega_2\zeta_2 \frac{\omega_2^2 + \omega^2}{2(\zeta_2^2\omega_2^2 + \omega^2)} + j\frac{\omega_2^2\omega - 2\omega_2^2\zeta_2^2\omega - \omega^3}{2(\zeta_2^2\omega_2^2 + \omega^2)}.
\end{aligned} \tag{4.25}$$

Ideal unit-step transient process achieved when imaginary part of the (4.25) equals to zero, which in turn gives

$$\omega = \omega_2 \sqrt{1 - 2\zeta_2^2}. \tag{4.26}$$

which apparently is the eigenfrequency of the secondary oscillations. However, as it has been mentioned earlier, better sensitivity is achieved when the sensitive element is driven in the primary resonance, which means

$$\omega = \omega_1 \sqrt{1 - 2\zeta_1^2}.$$

In this case

$$\omega_2 = \omega_1 \sqrt{\frac{1 - 2\zeta_1^2}{1 - 2\zeta_2^2}}. \tag{4.27}$$

Although this formula is somewhat different from the obtained earlier dependence (4.23), actual values are quite close. If the secondary natural frequency is chosen accordingly to (4.27) then the pole (4.25) becomes

$$s_1 = -\omega_1 \zeta_2 \sqrt{\frac{1 - 2\zeta_1^2}{1 - 2\zeta_2^2}} = -\omega_2 \zeta_2. \tag{4.28}$$

Obviously, unit-step settling time is still given by the expression (4.24), and all the hints to the settling time minimization remain the same.

Let us now demonstrate, by means of numerical simulations, how the suggested here choice of the excitation frequency and damping affects a unit-step angular rate transient process. Realistic numerical simulation is based on the approach described in Sect. 3.4. In addition to that, we would like to demonstrate how to run simulations using real and imaginary transfer functions (4.18) and to verify its performances in comparison to realistic sensitive element simulations. Simulink model, used to simulate CVG with real and imaginary transfer functions, is shown in Fig. 4.2.

Simulation results for these models are shown in Figs. 4.3 and 4.4, where solid line corresponds to the realistic model output, dotted line corresponds to the "realistic" reference output, and dashed line shows the input angular rate. Parameters of the simulations are as follows: $\omega_1 = 1000\pi$, $\omega_2 = 1.05\omega_1$, $\zeta_1 = \zeta_2 = 0.025$, $\omega = \omega_1$ for the non-optimised transient process, and $\omega_2 = 0.987\omega_1$ for the optimised in Fig. 4.4.

In the first case in Fig. 4.3, transient process for the non-optimised CVG is expressing significant overshoot and clear oscillatory behaviour. In the second case,

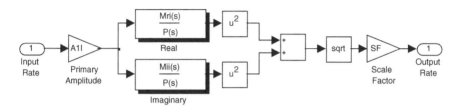

Fig. 4.2 Real and imaginary transfer functions model

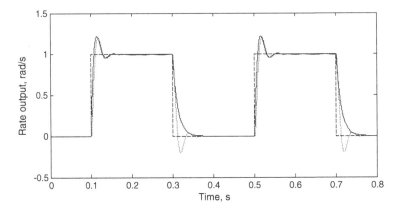

Fig. 4.3 Non-optimised transient process. (*solid* real and imaginary transfer functions model, *dashed* input angular rate, *dotted* realistic simulations output)

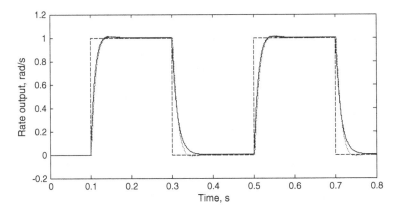

Fig. 4.4 Non-optimised transient process. (*solid* real and imaginary transfer functions model, *dashed* input angular rate, *dotted* realistic simulations output)

secondary natural frequency is chosen according to (4.23), which resulted in half-oscillation transient process as expected.

From the graphs in Figs. 4.3 and 4.4, one can also see that "realistic" transient process is somewhat different from the "complex" (based on real and imaginary transfer functions) one. This is believed to be the result of demodulation with the fixed phase shifted signal, while the actual phase shift varies in time. At the same time, "complex" output is much closer to the real secondary oscillations envelope, than the demodulated "realistic".

4.5 Simplified CVG Transfer Function and Its Accuracy

There is quite an important special case, when complex transfer functions transform to the simple real-valued one. Assuming equal primary and secondary natural frequencies ($\omega_1 = \omega_2 = k$), equal damping ratios ($\zeta_1 = \zeta_2 = \zeta$), resonance excitation ($\omega = k\sqrt{1 - 2\zeta^2}$), and constant angular rate, one can easily obtain

$$A_{20}(s) = \frac{g_2 q_{10} \sqrt{1 - 2\zeta^2}}{4\zeta k^2 (1 - \zeta^2)(s + k\zeta)} \Omega(s). \tag{4.29}$$

In this case, secondary amplitude (4.29) is related to the input angular rate by means of the following transfer function:

$$W_{20}(s) = \frac{A_{20}(s)}{\Omega(s)} = \frac{q_{10} g_2 \sqrt{1 - 2\zeta^2}}{4k^2 \zeta (1 - \zeta^2)(s + k\zeta)}. \tag{4.30}$$

As one can see, the simplified CVG transfer function (4.30) describes a simple first-order system with exponential (non-oscillatory) transient process.

When damping is small, e.g. $\zeta^2 \ll 1$, transfer function (4.30) can be rewritten as follows:

$$W_{20}(s) \approx \frac{q_{10} g_2}{4k^2 \zeta (s + k\zeta)}. \tag{4.31}$$

Finally, transfer function (4.31) relates angular rate to the secondary oscillations amplitude. However, more appropriate would be to consider transfer function relating unknown input angular rate to the measured angular rate, which can be easily obtained from (4.31) by dividing it on the steady state scale factor. The resulting transfer function is

$$W(s) = \frac{k\zeta}{s + k\zeta}. \tag{4.32}$$

Transfer function (4.32) represents a CVG system, shown in Fig. 4.5.

Here $\Omega^*(s)$ is the measured by CVG external angular rate.

Needless to say, that possibility to use transfer functions (4.30)–(4.32) for "non-tuned" CVG as well is highly desired. Therefore, let us evaluate accuracy of the function (4.32) in representing general case of CVG sensitive element dynamics.

In order to do that, let us compare transient processes produced by the simplified transfer function and by a numerical solution of Eq. (4.1) with subsequent demodulation. As a performance criterion the following integral function is used:

Fig. 4.5 CVG in control systems

Fig. 4.6 Integral error
of the transient process
representation

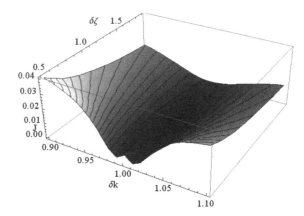

$$J(\delta k, \delta \zeta) = \int\limits_{0}^{T} [A_{20}(t) - A_{20}^{*}(t)]^2 dt. \qquad (4.33)$$

Here $\delta k = \omega_2/\omega_1$ is the ratio of the natural frequencies, $\delta \zeta = \zeta_2/\zeta_1$ is the ratio of the relative damping ratios, $A_{20}^{*}(t)$ is the demodulated secondary amplitude produced by the "realistic" model. Graphic plot of the functional (4.33) is shown below in Fig. 4.6.

Here, the central darker area corresponds to the perfectly tuned device ($\delta k = 1$, $\delta \zeta = 1$). One can see that wide range of sensitive elements with varying ratio of the natural frequencies and ratio of relative damping still could be represented by the transfer function (4.32) with acceptably low integral error (4.33).

4.6   Trajectory Rotation Transfer Function

In the previous chapter, it has been shown that in general case motion trajectory of the CVG sensitive element is an ellipse. Angle of the trajectory rotation in steady state is proportional to the angular rate. After simplified transfer functions of CVG sensitive element were derived, let us now derive transfer functions for trajectory angle of rotation and analyse corresponding transient processes.

As has been demonstrated in the previous section, Laplace transformation of the secondary amplitude with respect to settled primary oscillations and in case of the small damping is

$$A_{20}(s) = \frac{q_{10}g_2}{4k^2\zeta(s + k\zeta)}\Omega(s). \tag{4.34}$$

Constant real amplitude of the primary oscillations from (4.7) is

$$A_{10} = \frac{q_{10}}{\sqrt{(\omega_1^2 - \omega^2)^2 + 4\omega_1^2\zeta_1^2\omega^2}}, \tag{4.35}$$

Trajectory angle of rotation, given by (3.23) is

$$\theta = \frac{1}{2}\tan^{-1}\frac{2A_{10}A_{20}\cos\varphi}{A_{10}^2 - A_{20}^2},$$

where

$$\cos\varphi = \frac{2\zeta\delta k}{\sqrt{\delta k^4 - 2(1 - 2\zeta^2)\delta k^2 + 1}}.$$

Substituting expressions (4.34) and (4.35) into (3.23) results in the following expression for the trajectory angle of rotation in Laplace domain:

$$\theta(s) = \frac{1}{2}\tan^{-1}\left[\frac{4g_2k(s + k\zeta\delta k)\cos\varphi}{[4(s + k\zeta\delta k)^2 - g_2^2k^2\delta\Omega^2(s)]}\delta\Omega(s)\right]. \tag{4.36}$$

Here $\delta k = \omega_2/\omega_1$, $k = \omega_1$, $\delta\Omega(s) = \Omega(s)/\omega_1$. Apparently, expression (4.36) is non-linear in terms of the input angular rate. However, taking into account that relative angular rate is small ($\delta\Omega \ll 1$), expression (4.36) can be linearised with respect to the small $\delta\Omega$ as follows:

$$\theta(s) \approx \frac{g_2k\zeta\delta k}{(s + k\zeta\delta k)\sqrt{\delta k^4 - 2(1 - 2\zeta^2)\delta k^2 + 1}}\delta\Omega(s). \tag{4.37}$$

Finally, remembering that $\delta k = 1$ as was assumed for (4.34), expression (4.37) can be further simplified to

$$\theta(s) \approx \frac{g_2k}{2(s + k\zeta)}\delta\Omega(s). \tag{4.38}$$

Steady state of the obtained expression (4.38) is in a perfect agreement with the previously published steady state expressions for the motion trajectory angle of rotation (3.32).

Corresponding to (4.38) transfer functions from the relative angular rate to the trajectory rotation angle is as follows:

Fig. 4.7 CVG sensitive
element motion trajectory

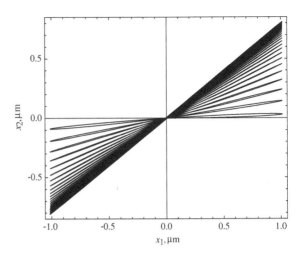

$$W_\theta^\Omega(s) = \frac{\theta(s)}{\delta\Omega(s)} = \frac{g_2 k}{2(s + k\zeta)}. \tag{4.39}$$

Transfer functions (4.39) can now be used to synthesise systems to control sensitive element motion trajectory as well as to implement advanced methods of the angular rate measurements.

Numerical simulation of the sensitive element motion trajectory based on Eq. (4.1) is shown in Fig. 4.7.

Primary oscillations are assumed to be already settled and constant angular rate is applied. Corresponding simulations for the angle of trajectory rotation are shown in Fig. 4.8.

Here dashed line corresponds to the simplified approximation (4.38). One can see that significant steady state error is present, which reduced usability of the derived simplified model (4.38).

Analysing expression (4.38) one can see that in steady state ($s = 0$) value of the θ angle is given by the simple ratio $g_2/2\zeta$. From the numerical simulation in Fig. 4.8 (dashed line) it is apparent that this value is not sufficiently accurate, while dynamic part appears to be acceptable. More accurate steady state value can be obtained directly from the expression (4.36), which results in the following improved approximation:

$$\theta(s) = \frac{1}{2}\tan^{-1}\left[\frac{4g_2\zeta\delta k}{\sqrt{\delta k^4 - 2(1 - 2\zeta^2)\delta k^2 + 1}}\right]\frac{\zeta k\delta k}{s + \zeta k\delta k}\delta\Omega(s). \tag{4.40}$$

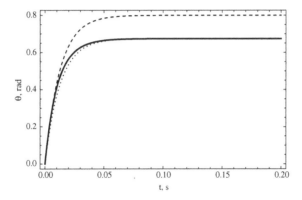

Fig. 4.8 Transient process simulations. (*solid* accurately simulated, *dashed* simplified approximation, *dotted* improved approximation)

Or in case of the sensitive element with matching natural frequencies

$$\theta(s) = \frac{1}{2}\tan^{-1}\left[\frac{4g_2}{\sqrt{2}}\right]\frac{\zeta k}{s + \zeta k}\delta\Omega(s). \tag{4.41}$$

Dotted line in Fig. 4.8 represents transient process simulated using (4.41). Apparently, these improved approximations are closer to the results of the direct simulations.

The following transfer function can be written similarly to (4.39):

$$W_\Omega^\theta(s) = \frac{\theta(s)}{\delta\Omega(s)} = \frac{1}{2}\tan^{-1}\left[\frac{4g_2}{\sqrt{2}}\right]\frac{\zeta k}{s + \zeta k}. \tag{4.42}$$

Obtained transfer functions for measured angular rate and for the angle of sensitive element trajectory rotation can now be used to develop signal processing systems as well as sensitive element control loops.

Resume

Mathematical model of CVG sensitive element in terms of demodulated signals led us to the important results, namely CVG transfer functions, where the external angular rate is no longer a coefficient, but an input to the system. More importantly, we can now analyse CVG sensitive element dynamics in terms of variable envelope amplitudes, instead of oscillatory primary and secondary motions. The simplicity of the obtained models allows efficient calculation and optimisation of main CVG performances, and, what is even more important, to synthesise signal processing and control systems using conventional methodologies.

Chapter 5
Sensitive Element Design Methodologies

Analysis of CVG-sensitive element motion equations and its solutions studied earlier allows us to derive expressions for major measurement performances and errors of CVG. This, in turn, makes possible to optimise performances by proper selection of the sensitive element parameters, as well as to reduce or even eliminate CVG measurement errors.

5.1 Optimal Excitation of the Primary Oscillations

Coriolis vibratory gyroscopes fabricated in miniature sizes by means of micromachining technologies, as well as many of the modern microelectromechanical systems (MEMS), use interdigitated microstructures both as an actuating and sensing component. A photograph of the typical interdigitated microstructure, which is part of the micromechanical gyroscope excitation system and often referred to as an electrostatic comb drive, is shown in Fig. 5.1.

Currently developed theories of the interdigitated microstructures that enable the analysis of non-linear effects are quite complicated. As a result, designers of MEMS still have the tendency to use numerical finite element method (FEM) simulations in order to model microsystems with electrostatic comb drives. Despite increased computational capabilities of the modern computers, complete FEM calculations still remain extremely time-consuming. Nevertheless, numerical approaches do not allow analytical analysis of the comb drive-based excitation systems. Hence the problem of creating a simple approximate approach to analyse the comb drive-based excitation systems to account for the essential for micromechanical gyroscope performances non-linear effects is addressed in this section.

Let us first consider operation of the primary oscillation excitation system. Circuit in Fig. 5.2 can describe general method of the resonator driving by means of a comb drive.

Here V_1 is the voltage applied to stators (fixed parts) of the comb drives 1 and 2, V_0 is the bias voltage applied to the inner moving mass, φ is the phase shift between voltages applied to the first and second comb drives. Total electrostatic force acting on the mass along the X-axis in this case can be determined as

© Springer International Publishing Switzerland 2016
V. Apostolyuk, *Coriolis Vibratory Gyroscopes*,
DOI 10.1007/978-3-319-22198-4_5

Fig. 5.1 Comb drive of
micromechanical gyroscope

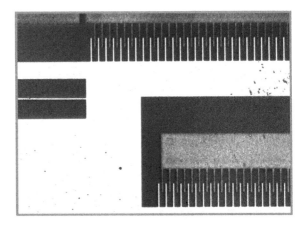

Fig. 5.2 Resonator driving
principle

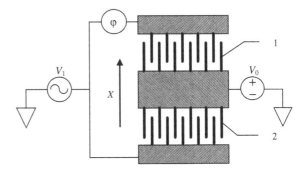

$$F_x = \frac{(V_1(\tau + \varphi) - V_0)^2}{2}\frac{\mathrm{d}C_1}{\mathrm{d}x} + \frac{(V_1(\tau) - V_0)^2}{2}\frac{\mathrm{d}C_2}{\mathrm{d}x}, \qquad (5.1)$$

where C_1 and C_2 are the capacitances of the comb drives 1 and 2 (Fig. 5.2) respectively, x is the displacement of the mass along the corresponding axis, τ is the phase of the driving voltage V_1. In case of the symmetrical and linear comb drives, where capacitances at the tips are negligible

$$\frac{\mathrm{d}C_1}{\mathrm{d}x} \approx -\frac{\mathrm{d}C_2}{\mathrm{d}x} = \frac{\mathrm{d}C}{\mathrm{d}x} \approx \mathrm{const},$$

and the force (5.1) will become

$$F_x = \frac{1}{2}[(V_1(\tau + \varphi) - V_0)^2 - (V_1(\tau) - V_0)^2]\frac{\mathrm{d}C}{\mathrm{d}x}. \qquad (5.2)$$

In micromechanical vibratory gyroscopes we usually want to have primary oscillations to have harmonic shape to make demodulation process as accurate as possible. We can assume therefore that $V_1 = V \sin(\omega t)$, $V_0 = V\delta V$. As a result, expression (5.2) becomes

$$F_x = \frac{V^2}{2}[(\sin(\omega t + \varphi) - \delta V)^2 - (\sin(\omega t) - \delta V)^2]\frac{dC}{dx}. \qquad (5.3)$$

It is apparent that the only parameter capable of affecting the shape of the excitation force (5.3) is the phase shift φ between voltages on the comb drives. Let us determine this phase shift from the maximum efficiency criterion. If the driving force does not depend on the displacement x, efficiency of the comb drive can be evaluated as

$$P(\delta V, \varphi) = \int\limits_0^{2\pi} [F_x(\tau)]^2 d\tau$$

$$= \frac{\pi}{2}(1 + 8\delta V^2 + \cos(\varphi)) \sin^2\left(\frac{\varphi}{2}\right) \cdot \left(V^2\frac{dC}{dx}\right)^2. \qquad (5.4)$$

Graphic plot of the numerically calculated performance criterion (5.4) is shown in Fig. 5.3.

Analysing the graph in Fig. 5.3, one should clearly see different optimal modes of excitation corresponding to different values for bias voltage δV. Let us identify these modes.

Maximum efficiency values for the phase shift φ and the voltage ratio δV as a parameter can be determined from the following equation

$$\frac{dP(\delta V, \varphi)}{d\varphi} = 0 \Rightarrow (4\delta V^2 + \cos\varphi)\sin\varphi = 0. \qquad (5.5)$$

Fig. 5.3 Comb drive efficiency

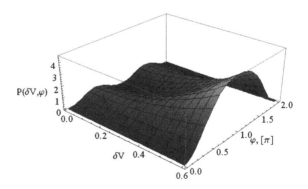

Solving Eq. (5.5) yields maximum efficiency phases given by the following equations:

$$\varphi = \arccos(-4\delta V^2), \quad \left(\delta V < \frac{1}{2}\right),$$

$$\varphi = \pi, \quad \left(\delta V \geq \frac{1}{2}\right), \tag{5.6}$$

$$\varphi = \frac{\pi}{2}, \quad (\delta V = 0).$$

Efficiency plot for the maximum efficiency modes given by the different bias voltages in expressions (5.6) is shown in Fig. 5.4.

It is apparent that there are two different optimal phase shifts for different values of the bias δV, leading to the two essentially different driving modes for the primary excitation: without bias voltage (grounded mass), and with bias, which is larger than a half of the driving voltage amplitude.

Total electrostatic forces acting on the mass in these modes will be determined by means of the following formula:

$$F_x = D(t)\frac{dC}{dx}. \tag{5.7}$$

Here $D(t)$ is the driving function, which is different for the two modes and can be calculated as

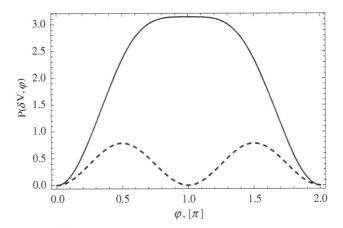

Fig. 5.4 Efficiency at different bias voltages (*solid* $\delta V = 0.5$, *dashed* $\delta V = 0$)

$$D(t) = 2V^2 \delta V \sin(\omega t), \quad \delta V \geq \frac{1}{2},$$

$$D(t) = \frac{V^2}{2} \cos(2\omega t), \quad \delta V = 0. \tag{5.8}$$

It has to be noted that "biased" mode ($\delta V \geq 1/2$) results in a larger driving force (higher efficiency) comparing with "grounded" mode ($\delta V = 0$). At the same time, driving force in the "grounded" mode will actuate with doubled frequency regarding to driving voltage frequency (see Fig. 5.5).

This effect of doubling frequency leads to possibility to separate excitation voltage from the sensing in the frequency domain. As a result, better signal to noise ratios can be achieved.

During all derivations presented above it was assumed that force does not depend on mass displacement. It means that $dC/dx \approx$ const, which is almost true for the small displacements. But in some applications of the comb drives it is necessary to achieve large displacement of the mass. The latter is often the case with the micromechanical gyroscopes, when higher amplitude of primary oscillations leads to higher sensitivity to the angular rate. In this case capacitance derivative is no longer constant and depends on displacements in a non-linear way.

Let us calculate capacitance for the comb structure cell that is shown in Fig. 5.6.

There are four basic capacitances in this structure: C_{ix} and C_{iy}—between stator ($i = 1$) and mass ($i = 2$) in the X and Y direction respectively. Other dimensions are L_i, B_i and H—length, width and height of the comb drive finger. Initial position of the mass in the shown reference system will be defined by means of four gaps G_{ix} and G_{iy}. If displacement of the mass relatively to stator will be defined by two variables x and y then corresponding capacitances will be

$$C_{1x} = \frac{\varepsilon\varepsilon_0 B_1 H}{G_{1x} - x}, \quad C_{2x} = \frac{\varepsilon\varepsilon_0 B_2 H}{G_{2x} - x},$$

$$C_{1y} = \frac{\varepsilon\varepsilon_0 (L_0 + x) H}{G_{1y} + y}, \quad C_{2y} = \frac{\varepsilon\varepsilon_0 (L_0 - x) H}{G_{2y} - y}. \tag{5.9}$$

Fig. 5.5 Driving force in different modes (*solid* "grounded" mode, *dashed* "biased" mode)

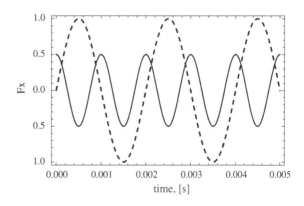

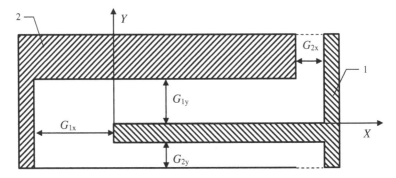

Fig. 5.6 Elementary comb drive cell

Here L_0 is the initial overlapping length, such as $L_0 = L_2 - G_{1x} = L_1 - G_{2x}$. Total capacitance between mass and stator will be a sum of all capacitances:

$$C(x,y) = C_{1x} + C_{2x} + C_{1y} + C_{2y}$$
$$= n\varepsilon\varepsilon_0 H\left(\frac{B_1}{G_{1x} - x} + \frac{B_2}{G_{2x} - x} + \frac{L_0 + x}{G_{1y} + y} + \frac{L_0 + x}{G_{2y} - y}\right). \qquad (5.10)$$

Here n is the total number of the elementary cells in the comb drive. Such capacitance will be no longer linear function of the displacements. Dependence of the capacitance from the displacements in x direction is presented in Fig. 5.7.

Apparently, for large displacements both along x and y coordinates we observe non-linearity in the capacitance, given by the expression (5.10). In Fig. 5.8 one can see the section of the graph in Fig. 5.7 along the x-axis.

Here dashed line corresponds to the "linear" capacitance, where tip widths B_1 and B_2 in (5.10) were neglected (set to zero). From this graph one can see that "linear" capacitance approximation is noticeably different from the actual one.

With respect to (5.10), capacitances C_1 and C_2 in (5.1) for the symmetrical comb drive will be

Fig. 5.7 Capacitance as a function of x and y displacements

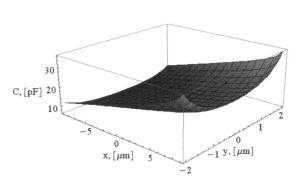

Fig. 5.8 Capacitance of the comb drive (*solid* non-linear, *dashed* neglecting tip width)

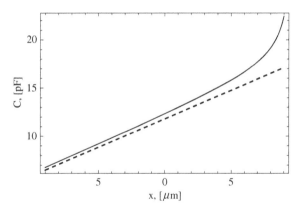

$$C_1(x, y) = C(x, y),$$
$$C_2(x, y) = C(-x, y),$$
$$\quad (5.11)$$

and hence correspondent derivatives for the force (5.1) can be represented as follows

$$\frac{dC_1}{dx} = Hn\varepsilon\varepsilon_0 \left[\frac{B_1}{(G_{1x} - x)^2} + \frac{B_2}{(G_{2x} - x)^2} + \frac{1}{G_{1y} + y} + \frac{1}{G_{2y} - y} \right],$$
$$\frac{dC_2}{dx} = -Hn\varepsilon\varepsilon_0 \left[\frac{B_1}{(G_{1x} + x)^2} + \frac{B_2}{(G_{2x} + x)^2} + \frac{1}{G_{1y} + y} + \frac{1}{G_{2y} - y} \right].$$
$$\quad (5.12)$$

Looking at the expressions (5.12) one can see that the displacements in the x direction will result in changes of the force in y direction as well. For some applications, such as comb driven micromechanical gyroscopes with double folded proof mass suspension, this will cause significant bias.

What else is important to note, is the absence of the overlapping length L_0 in the expression (5.11). This means, that actuation force does not depend on the overlapping length, while it is still present in the side force derivatives (5.12). From this point of view, introducing overlap diminishes influence of the primary oscillations on the quadrature mass motion.

Let us consider now force in the x direction for the non-linear capacitance comb drive. For the small displacements in x direction we can approximate derivatives in (5.11) by linear dependencies

$$\frac{dC_1}{dx} \approx nH\varepsilon\varepsilon_0 (a_0 + a_1 x),$$
$$\frac{dC_2}{dx} \approx -nH\varepsilon\varepsilon_0 (a_0 - a_1 x).$$
$$\quad (5.13)$$

Here the coefficients are given by the following expressions:

$$a_0 = \frac{B_1}{G_{1x}^2} + \frac{B_2}{G_{2x}^2} + \frac{1}{G_{1y}+y} + \frac{1}{G_{2y}-y},$$

$$a_1 = 2\left(\frac{B_1}{G_{1x}^3} + \frac{B_2}{G_{2x}^3}\right).$$

Thus net force acting on the mass in case of harmonic excitation with respect to (5.1) and (5.13) will be

$$F_x = \frac{nHV^2\varepsilon\varepsilon_0}{2}\Big[(a_0 + a_1 x)(\delta V - \sin(\omega t + \varphi))^2$$
$$-(a_0 - a_1 x)(\delta V - \sin(\omega t))^2\Big]. \tag{5.14}$$

Results of the efficiency analysis of the non-linear comb drive are similar to those considered earlier. The graph in Fig. 5.9 demonstrates accuracy of the approximation (5.13) in comparison with the more accurate expressions (5.11) with respect to the sum of the capacitance derivatives for both comb drives.

One should certainly note that the accuracy of the linear approximation is quite limited to the relatively small displacements.

Again there will be two optimal modes with the same phase shift. For the "biased" excitation mode

$$F_x = nHV^2\varepsilon\varepsilon_0[2a_0\delta V \sin(\omega t) + a_1(\delta V^2 + \sin^2(\omega t))x], \tag{5.15}$$

and for the "grounded" mode

$$F_x = \frac{nHV^2\varepsilon\varepsilon_0}{2}[a_1 x + a_0 \cos(2\omega t)]. \tag{5.16}$$

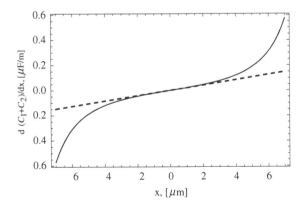

Fig. 5.9 Capacitance derivatives approximation (*solid* non-linear derivative, *dashed* linear approximation)

It has to be noted that for the biased excitation with large voltage ratio δV non-linearity of the capacitances will cause significant natural frequency shift that cannot be neglected for some applications. In order to reduce influence of the non-linear effects it is necessary to increase gaps in the x direction in comparison with the corresponding displacements.

Linear in terms of the displacement x force essentially perturbs the natural frequency of the primary oscillation mode. For the "grounded" mode force given by the (5.16), the natural frequency receives a constant shift as

$$k_* = \sqrt{k^2 - \frac{nHV^2\varepsilon\varepsilon_0 a_1}{2m}}, \qquad (5.17)$$

where k is the initial natural frequency related to the spring constant of the elastic suspension, m is the mass of the sensitive element. Actual force acting on the sensitive element is shown in Fig. 5.10.

In case of the "biased" mode force (5.15), the natural frequency will be variable in time:

$$k_* = \left[k^2 - \frac{nHV^2\varepsilon\varepsilon_0 a_1}{m}(\delta V^2 + \sin^2(\omega t))\right]^{\frac{1}{2}}. \qquad (5.18)$$

Actual force in significantly non-linear mode for the "biased" excitation mode is shown in Fig. 5.11.

Although forces in Figs. 5.10 and 5.11 are far from being harmonic, after they are applied to the spring–mass–damper system of the sensitive element, the resulting oscillations are not that much different from the harmonic shape. This is the result of the natural filtering properties of the oscillator. Nevertheless, if almost ideal harmonic excitation is desired, the gaps between tips of the combs G_{ix} must be

Fig. 5.10 Actual force acting on the sensitive element in the "grounded mode"

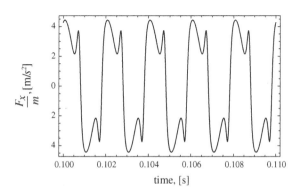

Fig. 5.11 Actual force acting
on the sensitive element in the
"biased mode"

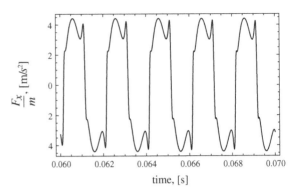

chosen 3–5 times larger than the expected amplitude of primary oscillations, and
excitation frequency must be adjusted according to the (5.17) and (5.18).

Developed above mathematical model of the comb drive-based excitation system allows analytical analysis of the micromechanical gyroscopes dynamics
without necessity for the time-consuming numerical simulations.

5.2   Scale Factor and Its Linearity

After we learned how to provide harmonic excitation to the primary mode of the
sensitive element, let us look at the secondary mode and its crucial characteristics
affecting measurement performances. When primary oscillations settled and occur
with constant amplitude, they provide a carrier signal that will be modulated with
the external angular rate to produce secondary oscillations. As follows from (3.15),
the amplitude of secondary oscillations is related to the angular rate

$$A_{20} = \frac{g_2 q_{10} \delta \omega}{|\Delta|} \delta \Omega \qquad (5.19)$$

where squared denominator is

$$\begin{aligned}
|\Delta|^2 &= k^8 \big[(1 - d_1 \delta\Omega^2 - \delta\omega^2)(\delta k^2 - d_2 \delta\Omega^2 - \delta\omega^2) \\
&\quad - \delta\omega^2 (g_1 g_2 \delta\Omega^2 + 4\zeta_1 \zeta_2 \delta k) \big]^2 \\
&\quad + 4k^8 \delta\omega^2 \big[\zeta_1 (\delta k^2 - d_2 \delta\Omega^2 - \delta\omega^2) + \zeta_2 \delta k (1 - d_1 \delta\Omega^2 - \delta\omega^2) \big]^2.
\end{aligned}$$

One could note that although amplitude of the secondary oscillations A_{20} appears
to be linearly related to the angular rate, it is actually not. Angular rate is also
present in the expression for the denominator in $|\Delta|$. While this non-linearity is

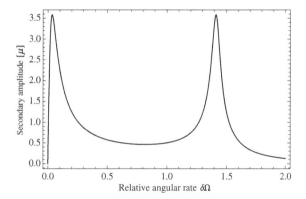

Fig. 5.12 Secondary amplitude as a function of angular rate

negligible for small angular rates, it starts to limit measurement range in case of high angular rates. Graphic plot of secondary amplitude as a function of angular rate is shown in Fig. 5.12 ($k = 1\,\text{Hz}$, $\zeta_1 = \zeta_2 = 0.025$, $\delta k = 1$, $\delta\omega = 1$).

Chosen in Fig. 5.12 range of angular rates is extremely wide and has no practical meaning (remember that angular rate is dimensionless and related to the primary natural frequency). However, this figure demonstrates how non-linear this dependence actually is. It is apparent that secondary amplitude A_{20} is close to being linearly related to the angular rate only for the very small angular rates ($\delta\Omega \ll 0.05$ as for the case in Fig. 5.12). This is demonstrated in Fig. 5.13.

In order to build a successful angular rate sensor, we need this response to be as linear as possible. In an ideal case, it should be related to the external angular rate as

$$A_{20}^* = S_\Omega \cdot \Omega, \qquad (5.20)$$

where A_{20}^* is the ideal CVG output, and S_Ω is the *scale factor*, which in this case relates angular rate to the ideal secondary amplitude, and is constant and depends on the sensitive element design parameters only.

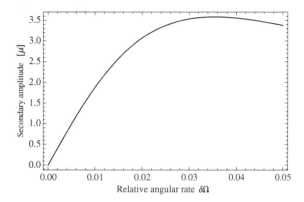

Fig. 5.13 Secondary amplitude for small angular rates

However, having looked at the expression (5.19) it is apparent that the actual scale factor depends on the angular rate as well. Ideal scale factor can be obtained from the expression (5.19) as a tangent to the actual dependence of the secondary amplitude taken at zero angular rate as

$$
\begin{aligned}
S_\Omega &= \left. \frac{\partial A_{20}}{\partial \Omega} \right|_{\Omega \to 0} \\
&= \frac{g_2 q_{10} \delta\omega}{k^3 \sqrt{\left[(1 - \delta\omega^2)^2 + 4\zeta_1^2 \delta\omega^2\right]\left[(\delta k^2 - \delta\omega^2)^2 + 4\zeta_2^2 \delta\omega^2 \delta k^2\right]}}
\end{aligned}
\tag{5.21}
$$

Scale factor (5.21) describes how sensitive is CVG to the angular rate. The higher the scale factor, the more CVG is sensitive to the angular rate. Analysing (5.21) one can see that scale factor depends on such sensitive element parameters as natural frequency ratio δk and relative driving frequency $\delta\omega$. Dependence of the scale factor from these parameters is shown in Fig. 5.14. The lighter the colour on the plot, the higher is scale factor. Apparently, maximum scale factor is achieved when all frequencies are perfectly matched, e.g. $\delta k = 1$ and $\delta\omega = 1$.

One can also clearly see two directions in Fig. 5.14 along which sensitivity is less dependent on parameters variations: $\delta\omega = 1$ and $\delta\omega = \delta k$. Sections of the surface from Fig. 5.14 along these two directions are shown in Fig. 5.15.

Moreover, taking into account cubic primary natural frequency k^3 in the denominator of (5.21), it is apparent that for better sensitivity natural frequency of primary oscillations has to be as low as possible. On the other hand, however, lowering primary natural frequency also reduces measurement range due to the non-linearity, as shown in Fig. 5.13.

Fig. 5.14 CVG scale factor as a function of design parameters

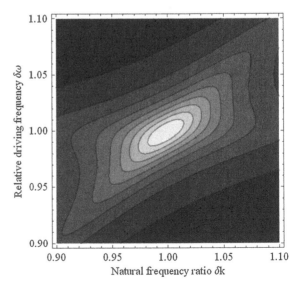

Fig. 5.15 Scale factor as a
function of natural frequency
ratio $\delta\omega$ (*solid* $\delta\omega = 1$,
dashed $\delta\omega = \delta k$)

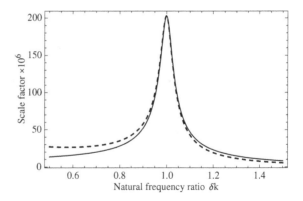

In order to properly account for the scale factor non-linearity, let us introduce a
dimensionless non-linearity factor as

$$L = 1 - \frac{A_{20}}{A_{20}^*}. \tag{5.22}$$

Graphic plot of the non-linearity factor (5.22), expressed in percents, as a
function of the relative angular rate is shown in Fig. 5.16.

Assuming certain maximum acceptable value for non-linearity (5.22), it is
possible to find such an angular rate, at which this non-linearity is reached. The
following approximate expression can be used to calculate this relative angular rate:

$$\delta\Omega^* = \left\{ \frac{L_0\left[(\delta k^2 - \delta\omega^2)^2 + 4\delta k^2\delta\omega^2\zeta_2^2\right] \cdot \left[(1 - \delta\omega^2)^2 + 4\delta\omega^2\zeta_1^2\right]}{(\delta\omega^2 - 1)D_0 + 4\delta\omega^2\left[g_1 g_2\delta k\delta\omega^2\zeta_1\zeta_2 - d_1\zeta_1^2(\delta k^2 - \delta\omega^2)\right]} \right\}^{\frac{1}{2}}. \tag{5.23}$$

Fig. 5.16 CVG non-linearity
as a function of angular rate
(*solid* non-linearity factor,
dotted approximation, *dashed
line* corresponds to 1 %
non-linearity)

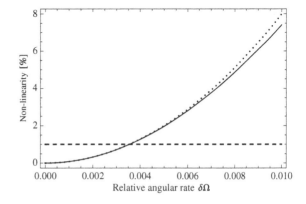

Here

$$D_0 = \left(\delta k^2 - \delta \omega^2\right)\left(d_1 + d_2\delta k^2 - (d_1 + d_2 - g_1g_2)\delta \omega^2\right) + 4d_2\delta k^2\delta \omega^2\zeta_2^2,$$

and L_0 is the assumed acceptable level of non-linearity.

Expression (5.23) can be further simplified if driving occurs at the primary natural frequency, e.g. $\delta \omega = 1$:

$$\delta \Omega^* = \left\{ \frac{L_0\zeta_2\left[1 + \delta k^4 - 2\delta k^2(1 - 2\zeta_1^2)\right]}{g_1g_2\delta k\zeta_1 - d_1\zeta_2(\delta k^2 - 1)} \right\}^{\frac{1}{2}}. \tag{5.24}$$

Thus, when an acceptable level of non-linearity is chosen along with the measurement range, minimum possible primary natural frequency can be calculated using (5.23) or (5.24) as

$$k_{\min} = \frac{\Omega_{\max}}{\delta \Omega^*}. \tag{5.25}$$

For example, if $L_\Omega = 0.01$ (or 1 %) and $\Omega_{\max} = 1\,\text{Hz}$ then minimal value for the natural frequency of primary oscillations will be $k_{\min} \approx 281\,\text{Hz}$. Although this value is relatively low, implying that the lower limit will, in fact, be determined by other factors, but nevertheless there is no reason to design CVG sensitive element with higher primary frequency, since it is reducing its sensitivity.

5.3   Resolution and Dynamic Range

There are many different techniques used to detect secondary displacements of the proof mass. Among them are capacitive, piezoresistive, piezoelectric, magnetic, optical, etc. Needless to say that the simplest to implement and the most widely spread among micromechanical devices is, of course, capacitive. Assuming that one uses capacitive detection of the secondary oscillations, *resolution* of the Coriolis vibratory gyroscope can be obtained be means of given minimal capacitance changes, which the system is capable of detecting. Let us denote this minimal change of capacitance as $\Delta C_{\min}$. Since capacitance C is a function of proof mass displacement x, we can write

$$C(x) = C(0) + \frac{\partial C(0)}{\partial x}x + O(x^2).$$

For the small displacements x, which is true for the secondary oscillations, we can neglect the $O(x^2)$ terms and the capacitance change will be given as

$$\Delta C(x) = C(x) - C(0) \approx \frac{dC(0)}{dx} x. \tag{5.26}$$

In case of differential measurement, which are quite commonly accepted in capacitance measurements, the resulting capacitance change is produced by subtraction of two separately measured capacitances C_1 and C_2 as follows:

$$\Delta C(x) = C_1(x) - C_2(x) \approx 2\frac{dC(0)}{dx} x. \tag{5.27}$$

Change in capacitance of two parallel conductive plates, for instance, caused by displacements of the proof mass in case of differential measurement (5.27) can be calculated as

$$\Delta C = \frac{\varepsilon \varepsilon_0 S}{x_0 - x} - \frac{\varepsilon \varepsilon_0 S}{x_0 + x} \approx 2\frac{\varepsilon \varepsilon_0 S}{x_0^2} x. \tag{5.28}$$

Here, x_0 is the base gap between the electrodes, x is the displacement of the electrodes, S is the overlapped area, ε is the relative dielectric constant of the proof mass environment and ε_0 is the absolute dielectric constant of vacuum. The shift of the electrodes caused by changes of the angular rate $\Delta\Omega$ is given by

$$x = r_0 S_\Omega \Delta\Omega, \tag{5.29}$$

where S_Ω is determined by expression (5.21), r_0 is the distance from the rotation axis to the centre of electrode for the rotary sensitive element and unity for the translational sensitive element.

Thus, combining Eqs. (5.28) and (5.29), we can obtain the resolution of a single mass micromechanical vibratory gyroscope that is given by

$$\Delta\Omega_{min} = \frac{\Delta C_{min} k^3 \sqrt{\left(\left(\delta k^2 - \delta\omega^2\right)^2 + 4\delta k^2 \delta\omega^2 \zeta_2^2\right)\left(\left(1 - \delta\omega^2\right)^2 + 4\delta\omega^2 \zeta_1^2\right)}}{2\frac{dC(0)}{dx} r_0 g_1 q_2 \delta\omega}. \tag{5.30}$$

Here the best resolution corresponds to a minimal $\Delta\Omega_{min}$. Note that formula (5.30) represents the resolution with a capacitive differential readout. However, the same procedure can be applied to any readout principle using expression (5.21) for the scale factor. The resolution, which is given by formula (5.30), is related to the dynamics of the sensitive element and is fundamental from the design point of view. The real resolution of the gyroscope cannot be better than the one determined by the dynamics of its sensitive element. Unfortunately, the resolution can be worse since it is also affected by noise.

The resolution alone would never give to the user complete understanding of the measuring capabilities of a CVG gyroscope since it is tightly linked to the measurement range. The same resolution over different measurement ranges will correspond to gyroscopes with entirely different performances. Therefore, another characteristic is widely used to describe measuring capabilities of sensors, namely dynamic range, which in case of an angular rate sensor is defined as follows:

$$R = 20 \log_{10} \frac{\Omega_{\text{max}} - \Delta\Omega_{\text{min}}}{\Delta\Omega_{\text{min}}}. \tag{5.31}$$

Here the dynamic range R is expressed in dB, $\Delta\Omega_{\text{min}}$ is given by expression (5.30), assuming that the sensor threshold is equal to its resolution, Ω_{max} is the maximum angular rate that can be measured with acceptable errors, which can be calculated from (5.25) as

$$\Omega_{\text{max}} = \delta\Omega^* k, \tag{5.32}$$

where k is the natural frequency of the primary oscillations. Graphic plot of the dynamic range as a function of the primary natural frequency is shown in Fig. 5.17.

Looking at Fig. 5.17, one can see that the lower the primary natural frequency is the higher will be dynamic range of a CVG. Needless to say that for any reasonable required dynamic range the corresponding sensitive element can be designed even without vacuum packaging (solid line in Fig. 5.17). Despite this obvious fact,

Fig. 5.17 Dynamic range as a function of the primary natural frequency (*solid* $\zeta_1 = \zeta_2 = 0.025$, *dashed* $\zeta_1 = \zeta_2 = 0.0025$, $\delta\omega = \delta k = 1$)

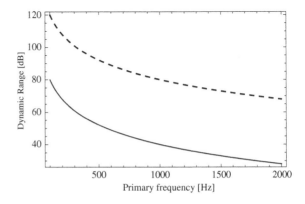

micromechanical CVGs are still referred to as a low-grade angular rate sensor. The reason for that is usage of micromachining for the fabrication of the gyroscopes in particular and the approach towards development of "miniature" sensors in general. As soon as designers try to develop a "micromechanical" gyroscope they make it extremely small in size, comparing to the conventional angular rate sensors. The overall size of the sensitive element in every direction varies from 100 to 5000 μm. As a result, the natural frequency of the primary oscillations ends up in a range from 5 to 100 kHz. Apparently, in order to produce any, not mentioning high grade, angular rate sensing with such devices extremely high vacuum packaging is necessary. On the other hand, if one will try to design a gyroscope with low primary frequency, this will require making huge proof mass and very thin and long springs of the elastic suspension. This is quite a complicated task if micromachining is used, especially considering very high relative tolerances of this fabrication process.

5.4 Bias

Bias in micromechanical gyroscopes can be the result of many different factors. Let us consider sources of bias concerned with the sensitive element and its dynamics. One of these is vibration at the excitation frequency. The interference of vibrations at other frequencies will be small and can be filtered. It is obvious that for the translational gyroscopes, only translational vibration will have an effect, and for rotational gyroscopes only angular vibrations will be relevant. Therefore, in the case of vibrations at operation frequency, the motion equations of the sensitive element will be

$$\begin{cases} \ddot{x}_1 + 2\zeta_1\omega_1\dot{x}_1 + (\omega_1^2 - d_1\Omega_3^2)x_1 + g_1\Omega_3\dot{x}_2 = q_1(t) + w_1(t), \\ \ddot{x}_2 + 2\zeta_2\omega_2\dot{x}_2 + (\omega_2^2 - d_2\Omega_3^2)x_2 - g_2\Omega_3\dot{x}_1 = w_2(t). \end{cases} \tag{5.33}$$

Here $w_1(t)$ and $w_2(t)$ are components of the acceleration vector that represents the motion of the base reference system. By representing the vibrations as $w_i = w_{i0}\cos(\omega t)$, we can obtain the solution on the amplitude of secondary oscillations in dimensionless form

$$A_{W2} = \frac{g_2 q_1 \delta\omega\delta\Omega + \sqrt{w_{20}^2(1 - \delta\Omega^2 - \delta\omega^2)^2 + \delta\omega^2(2\zeta_1 w_{20} + g_2\delta\Omega w_{10})^2}}{k^2\Delta},$$

$$\tag{5.34}$$

If we denote the amplitude without vibrations as A_{20}, which is given by (5.19), then the relative error caused by vibration at excitation frequency is given by

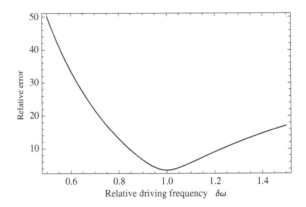

Fig. 5.18 Typical error from vibrations as a function of relative driving frequency ($\zeta_1 = \zeta_2 = 0.025$, $q_{10} = w_1 = w_2 = 10$, $\delta\Omega = 0.01$)

$$\delta A_W = \frac{A_{W2} - A_{20}}{A_{20}}$$
$$= \frac{\sqrt{w_{20}^2\left(1 - d_1\delta\Omega^2 - \delta\omega^2\right)^2 + \delta\omega^2(2\zeta_1 w_{20} + g_2\delta\Omega w_{10})^2}}{g_2 q_1 \delta\omega\delta\Omega}. \qquad (5.35)$$

Let us note that the error arising from vibration does not depend on the ratio between the natural frequencies but depends on the relative drive frequency. This dependency is shown in Fig. 5.18.

It can easily be proven that the minimal value for this error achievable at driving frequency that is a solution of the following equation:

$$1 - \delta\omega^2 - d_2\delta\Omega^2 = 0 \Rightarrow \delta\omega = \sqrt{1 - d_2\delta\Omega^2} \approx 1. \qquad (5.36)$$

This result also ensures that it is preferable to drive the primary oscillations at their resonance.

Another source of bias is a misalignment between elastic and readout axes. This is most typical for the translation sensitive elements. The linearised motion equations in this case will be as follows:

$$\begin{cases} \ddot{x}_1 + 2\zeta_1\omega_1\dot{x}_1 + \left(\omega_1^2 - d_1\Omega^2\right)x_1 + g_1\Omega\dot{x}_2 - 2\theta\Delta\omega_1^2 x_2 = q_1(t), \\ \ddot{x}_2 + 2\zeta_2\omega_2\dot{x}_2 + \left(\omega_2^2 - d_2\Omega^2\right)x_2 - g_2\Omega\dot{x}_1 + 2\theta\Delta\omega_2^2 x_1 = 0. \end{cases} \qquad (5.37)$$

Here θ is the misalignment angle, $\Delta\omega_2^2 = (k_2 - k_1)/2M_2$, $\Delta\omega_1^2 = (k_1 - k_2)/2M_1$, where k_1 and k_2 are stiffness coefficients, corresponding to primary and secondary oscillations respectively, M_1 and M_2 are inertia factors that for translational motion $M_1 = m_1 + m_2$, $M_2 = m_2$, and for rotational motion

$M_1 = I_{11} + I_{22}$, $M_2 = I_{22}$. The amplitude of the secondary oscillations in this case will be

$$A_{20} = \frac{q_{10}\sqrt{g_2^2\delta\omega^2\delta\Omega^2 + 4\theta^2\delta\Delta\omega_2^4}}{\omega_0^2\Delta_\theta}, \qquad (5.38)$$

where

$$
\begin{aligned}
\Delta_\theta^2 = & \left[\left(\delta k^2 - d_2\delta\Omega^2 - \delta\omega^2\right)\left(1 - d_1\delta\Omega^2 - \delta\omega^2\right)\right. \\
& -\delta\omega^2\left(4\delta k\zeta_1\zeta_2 + g_1g_2\delta\Omega^2\right)\Big]^2 \\
& + 4\delta\omega^2\left[\delta k\zeta_2\left(1 - d_1\delta\Omega^2 - \delta\omega^2\right) + \zeta_1\left(\delta k^2 - d_2\delta\Omega^2 - \delta\omega^2\right)\right. \\
& \left.- 2\delta\Omega\theta\left(\delta\Delta\omega_1^2 + \delta\Delta\omega_2^2\right)\right]^2.
\end{aligned}
$$

It is apparent that if $\theta = 0$ then there is no error arising from misalignment. Moreover, this error will also be absent in the following case

$$\Delta\omega_2^2 = \frac{k_1 - k_2}{2m_2} = 0 \Rightarrow k_1 = k_2. \qquad (5.39)$$

Here k_i are the stiffness coefficients of the elastic suspension and m_1 is the effective mass of secondary oscillations. In addition, we can represent the amplitude (5.39) as a sum of two components, namely, one arising from the angular rate and the other caused by misalignment. In this case we can determine the relative error from such misalignment as

$$\delta A_\theta = \frac{A_\theta}{A_{20}} = \frac{\theta^2\delta\Delta\omega_2^4}{g_2\delta\omega^2\delta\Omega^2}, \quad (\Omega \neq 0). \qquad (5.40)$$

On the other hand, we can find an acceptable tolerance for the misalignment θ_{max} with respect to the given acceptable relative bias $\delta\Omega_{max}$ and under the condition of no rotation as

$$\theta_{max} = \frac{\delta\Omega_{max}\delta\omega}{\delta\Delta\omega_2^2}. \qquad (5.41)$$

Formula (5.41) also gives us an angle of misalignment if bias is known. This value can be used for algorithmic bias compensation. If we can obtain information about external accelerations at the operation frequency the bias can be compensated based on dependence (5.35).

A generalised approach to compensation of bias caused by elastic and damping alike cross-coupling using signal processing will be described in the next chapter.

5.5 Dynamic Error and Bandwidth

By definition, dynamic error is the error in angular rate measurements due to varying in time angular rate. To simplify dynamic error analysis, angular rate is usually assumed to be harmonic, e.g. oscillating in time with some frequency λ. Dynamic error of CVG can by defined in terms of the amplitude and phase responses to this harmonic angular rate. Expressions for these responses were derived in the previous chapter and are

$$A(\lambda) = \frac{q_{10}(d_4\lambda + g_2\omega)}{\sqrt{[(\omega_2^2 - (\lambda + \omega)^2)^2 + 4\zeta_2^2\omega_2^2(\lambda + \omega)^2][(\omega_1^2 - \omega^2)^2 + 4\zeta_1^2\omega_1^2\omega^2]}},$$

$$\varphi(\lambda) = \tan^{-1}\left\{\frac{[\omega_2^2 - (\lambda + \omega)^2][\omega_1^2 - \omega^2] - 4\omega_1\omega_2\zeta_1\zeta_2\omega(\lambda + \omega)}{2[\omega_2\zeta_2(\lambda + \omega)(\omega_1^2 - \omega^2) + \omega_1\zeta_1\omega(\omega_2^2 - (\lambda + \omega)^2)]}\right\}.$$

$$(5.42)$$

In an ideal case, amplitude and phase of the secondary oscillations for the harmonic angular rate must by the same as for the constant one. This allows defining the dynamic error both for the amplitude and phasing as follows:

$$E_A = \frac{A(\delta\lambda)}{A(0)},$$

$$E_\varphi = \frac{\varphi(\delta\lambda)}{\varphi(0)}.$$

$$(5.43)$$

Here $\delta\lambda = \lambda/\omega_1$ is the relative frequency of angular rate oscillations. Errors (5.43) are dimensionless and are equal to 1 in an ideal case. More importantly, dynamic errors allow to define important CVG performance, which is *bandwidth*, as a range of angular rate frequencies, where dynamic errors are within given tolerance.

Let us first study the phase dynamic error. Except of the relative angular rate frequency, phase dynamic error depends on such design parameters of the sensitive element as relative excitation frequency $\delta\omega = \omega/\omega_1$, natural frequency ratio $\delta k = \omega_2/\omega_1$, relative damping ratio $\delta\zeta = \zeta_1/\zeta_2$, and damping factor of the primary oscillations $\zeta = \zeta_1$. As has been shown earlier, it is advantageous to excite the sensitive element at the natural frequency of primary oscillations ($\delta\omega = 1$). In this case graphic plot of the phase dynamic error for different primary damping factors is shown in Fig. 5.19.

Analysis of this graph suggests that although it appears that the best case is the absence of damping at all, the presence of even a small amount of damping significantly increases the phase dynamic error. At the same time, increasing the damping causes the error to approach the ideal case.

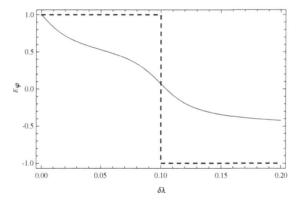

Fig. 5.19 Phase dynamic error (*solid line* $\zeta = 0.02$, *dashed line* $\zeta = 0$, $\delta k = 1.1$, $\delta \omega = 1$, $\delta \zeta = 1$)

Let us now study the amplitude dynamic error. Substituting the amplitude from (5.42) into the (5.43) gives the expression for the amplitude dynamic error:

$$E_A(\delta\lambda) = \left\{ (\delta\lambda + g_2\delta\omega)[(\delta k^2 - \delta\omega^2)^2 + 4\delta k^2\zeta^2\delta\zeta^2\delta\omega^2]^{1/2} \right.$$
$$\times [(1 - \delta\omega^2)^2 + 4\zeta^2\delta\omega^2]^{1/2} \Big\} \Big/ \Big\{ g_2\delta\omega[(\delta k^2 - (\delta\lambda + \delta\omega)^2)^2$$
$$\left. + 4\delta k^2\zeta^2\delta\zeta^2(\delta\lambda + \delta\omega)^2]^{1/2}[(1 - (\delta\lambda + \delta\omega)^2)^2 + 4\zeta^2(\delta\lambda + \delta\omega)^2]^{1/2} \right\}$$

$$(5.44)$$

Note that amplitude dynamic error, given by (5.44), does not explicitly depend on the primary natural frequency. Maximisation of CVG sensitivity requires small natural frequency of the primary oscillations, due to the k^3 term in the denominator of the scale factor (5.21). Providing the necessary bandwidth of the sensor requires keeping the amplitude dynamic error as low as possible (ideally equal to 1) within that bandwidth. Graphic plot of the amplitude dynamic error as function of the primary oscillations damping coefficient ζ and relative frequency of the angular rate is shown in Fig. 5.20.

From this graph one could see that decreased damping results in significant dynamic error even for the small frequencies of the angular rate. On the other hand, increased damping causes the drop between peaks to diminish, thus providing low values for the dynamic error.

Amplitude dynamic error has two maximums and one local minimum along the rate frequency axis that are clearly visible on Fig. 5.20, especially in the case of low damping. Positions of these extremes can be found from the following equation:

$$\frac{d}{d(\delta\lambda)} E_A(\delta\lambda) = 0. \qquad (5.45)$$

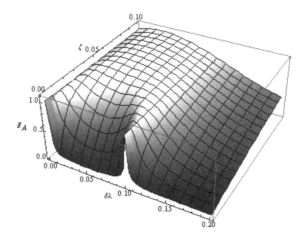

Fig. 5.20 Amplitude dynamic error ($\delta k = 1.1$, $\delta \omega = 1$, $\delta \zeta = 1$)

General solution of Eq. (5.45) is quite difficult to analyse. However, in case of zero damping it can be significantly simplified and its solutions could be found from the following equation:

$$(\delta\lambda + \delta\omega)[1 + \delta k^2 - 2(\delta\lambda + \delta\omega)^2][\delta k^2 - (\delta\lambda + \delta\omega)^2][(\delta\lambda + \delta\omega)^2 - 1] = 0.$$
(5.46)

Three positive roots of Eq. (5.46) are

$$\delta\lambda_1 = 1 - \delta\omega,$$

$$\delta\lambda_2 = \sqrt{\frac{1 + \delta k^2}{2}} - \delta\omega,$$
(5.47)

$$\delta\lambda_3 = \delta k - \delta\omega.$$

Here the first and the last roots correspond to maximums, and the second one to the minimum.

In general terms, optimization of the bandwidth means providing the same ideal level of the amplitude dynamic error at each of three extremes in vicinity of the given by (5.47) frequencies of the angular rate. Amplitude dynamic error level at the second maximum, which corresponds to $\delta\lambda_3$, can be controlled by the damping ratio $\delta\zeta$, provided it is a root of the following equation:

$$E_A(\delta\lambda_3) = 1.$$
(5.48)

Positive solution of Eq. (5.48), assuming $\delta\omega = 1$, is

$$\delta\zeta = (\delta k^2 - 1)(\delta k + g_2 - 1)/\delta k[g_2^2(\delta k^2 + \delta k^6$$
$$- 4\zeta^2 + 2\delta k^4(2\zeta^2 - 1)) - 4\zeta^2(\delta k - 1)^2 - 8g_2\zeta^2(\delta k - 1)]^{1/2}. \qquad (5.49)$$

Now let us find the damping ζ that satisfies the equation

$$E_A(\delta\lambda_2) = 1, \qquad (5.50)$$

where $\delta\zeta$ is given by the expression (5.49). In this case Eq. (5.50) will depend only on natural frequencies ratio and unknown damping ζ. Full expression for Eq. (5.50) is quite large to be shown here, however, it could be easily solved numerically. Graphic plot of the amplitude dynamic error as a function of damping ζ and with respect to the optimal damping ratio (5.49) is shown in Fig. 5.21.

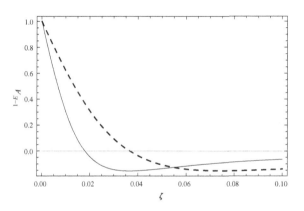

Fig. 5.21 Minimal dynamic error damping (*solid* $\delta k = 1.05$, *dashed* $\delta k = 1.1$)

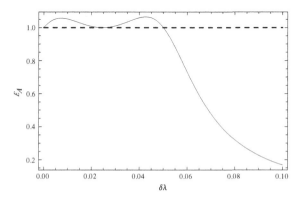

Fig. 5.22 Optimised amplitude dynamic error

For example, optimal damping parameters for $\delta k = 1.05$, are $\zeta = 0.018$, $\delta\zeta = 0.921$. Amplitude dynamic error for this case is shown in Fig. 5.22.

One should also note that achieved level of the amplitude dynamic error could be further improved if the objective level of the dynamic error in Eqs. (5.48) and (5.50) is set to $1 - e$ instead of 1, where e is the acceptable value of the dynamic error (dynamic error tolerance).

Based on the presented above analysis of the dynamic errors of the CVGs, necessary bandwidth can be achieved by means of proper choice not only of the natural frequencies ratio, which affects position of the second peak according to (5.47), but by providing proper damping of the primary and secondary oscillations as well.

In order to provide necessary bandwidth, natural frequencies ratio could be chosen based on the position of the second maximum in the amplitude response:

$$\delta k = \delta\lambda_* + \delta\omega. \tag{5.51}$$

Here $\delta\lambda_*$ is the required bandwidth in the dimensionless form, related to the natural frequency of primary oscillations. After the natural frequency ratio is calculated using (5.51), the result is used to calculate necessary damping for the primary oscillations. The latter problem can be solved either numerically or even analytically for some simplified cases. Having now calculated proper frequency ratio and primary damping, corresponding secondary damping is calculated using the ratio (5.49).

Considering the fact that providing necessary damping in the CVGs is not an easy task, optimal values can be implemented via creating "electrical damping" by using closed-loop feedback control both for the primary and secondary oscillations.

Resume

In this chapter we covered calculation and optimisation of main CVG performances. It is always beneficial to choose parameters of a sensitive element based on direct calculations by given dependencies, than finding them by costly, not mentioning requiring a lot of efforts, trial and error in prototyping. One could note that there is no universal recipe to design a perfect CVG-sensitive element. It is always a balance between different trade-offs, when one performance is sacrificed in favour of another. And being able to estimate the results of such trade-offs tremendously improves the design process.

Chapter 6
Signal Processing and Control

Performances of Coriolis vibratory gyroscopes can be improved not only by proper sensitive element design, but by subsequent signal processing as well. Noises and disturbances can be either completely filtered out or reduced by means of static or dynamic filtering and well-modelled errors can be compensated using both modulated and demodulated signal processing and control. In this chapter, we study different algorithms of signal processing aimed at the improvement of CVG performances.

6.1 Process and Sensor Noises in CVG

Performances of CVG can be affected by uncontrolled stochastic influences in two ways: as a "sensor noise", which is added to the output of the system, and as a "process noise" or disturbances, which are added to the input of the system. The latter could be also treated as "rate-like" disturbances. Figure 6.1 demonstrates how these two noises are influencing the CVG system.

Here $W(s)$ is the system transfer function that represents the CVG-sensitive element dynamics which is studied in detail earlier in Chap. 4:

$$W(s) = \frac{k\zeta}{s + k\zeta} \tag{6.1}$$

and $G(s)$ represents some filter, improving CVG performances, ψ is the stochastic disturbance and φ is the sensor noise. Ω_0 is the actual angular rate and Ω is the angular rate measured by CVG.

Sensor and process noises are usually modelled as different stochastic processes given by its power spectral densities.

One of the most widely used models is the Gaussian white noise that is described by the following constant power spectral density expressed in Laplace domain:

$$S_w(s) = \sigma^2. \tag{6.2}$$

© Springer International Publishing Switzerland 2016
V. Apostolyuk, *Coriolis Vibratory Gyroscopes*,
DOI 10.1007/978-3-319-22198-4_6

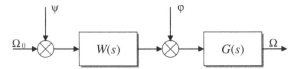

Fig. 6.1 Process and sensor noises in CVG

This stochastic process has zero mean and σ is its standard deviation. In order to relate noise power to the power of the angular rate in our later derivations, the noise-to-signal ratio γ can be added to (6.2) as

$$S_w(s) = \gamma^2 \sigma^2. \tag{6.3}$$

Another widely used noise representation is Brownian noise or random walk noise, which is defined as an integral of the white noise. It has the following power spectral density:

$$S_b(s) = \frac{\gamma^2 \sigma^2}{-s^2}. \tag{6.4}$$

In case when noise is present only in a certain bandwidth, it can be modelled by the following low-pass spectral density

$$S_l(s) = \frac{\gamma^2 \sigma^2 B^2}{B^2 - s^2}, \tag{6.5}$$

where B is the cut-off frequency, or by its complimentary high-pass spectral density

$$S_h(s) = -\frac{\gamma^2 \sigma^2 s^2}{B^2 - s^2}. \tag{6.6}$$

Spectral densities (6.2)–(6.6) can now be used to synthesise different filters that reduce its influence on CVG performances.

6.2 Sensor Noise Optimal Filtering

There are essentially two kinds of optimal filtering approaches that can be used to reduce the effect of the sensor noise in CVG. One is static filtering and the other one is dynamic. The former means that fixed structure filter is designed that is represented by its transfer function, and this function does not vary in time. Optimal static filters are derived using the Wiener approach. Dynamic filtering, on the contrary, is an algorithm that runs on a microcomputer and can adjust its parameters and/or structure to achieve maximum filtering performance. As an example, one could consider the digital Kalman filter which can be effectively used to remove sensor noise in CVG. Both of these approaches have their advantages and short-comings. For instance, static filter does not require any microcomputer and can be implemented within an application-specific integrated circuit (ASIC) that is fabri-cated along with the sensitive element. At the same time, dynamic filtering can provide better and adjustable performances.

Let us now consider a static filter attached to the output of CVG system as shown in the Fig. 6.2.

Here $G(s)$ is the optimal static filter transfer function, producing output x, which is filtered from the sensor noise φ measured angular rate Ω.

General algorithm of the optimal filter synthesis for the system in Fig. 6.2 has been developed by Wiener for the stationary stochastic sensor noise and is as follows.

Error of the system is defined as a difference between the actual output of the system Ω and the ideal output which is the given by the desired transformation $H(s)$ of the input as

$$\varepsilon = \Omega - H(s) \cdot \Omega_0. \tag{6.7}$$

It is also assumed that signals Ω and Ω_0 are the centred (zero mean) stochastic processes defined in terms of system transfer functions and known spectral densities of the input angular rate and sensors noises.

Performance criterion for the system is assumed to be in the form of the fol-lowing functional:

$$J = E\{\varepsilon' \cdot \varepsilon\} = \frac{1}{j} \int_{-j\infty}^{j\infty} S_{\varepsilon\varepsilon}(s) ds. \tag{6.8}$$

Here $S_{\varepsilon\varepsilon}(s)$ is the error spectral density which can be calculated from the system transfer functions and signal spectral densities using Wiener–Khinchin theorem as follows:

$$S_{\varepsilon\varepsilon}(s) = (GW - H)S_\Omega(W_* G_* - H_*) + (GW - H)S_{\varphi\Omega}G_* \\ + GS_{\Omega\varphi}(W_* G_* - H_*) + GS_\varphi G_*, \tag{6.9}$$

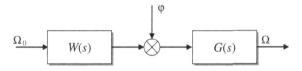

Fig. 6.2 Static sensor noise filtering

where asterisk designates complex conjugate, $S_{\Omega\varphi}(s)$ and $S_{\varphi\Omega}(s)$ are the cross-spectral densities between input angular rate and additive sensor noise which in general are assumed to be known.

By means of introducing new variables defined as

$$
\begin{aligned}
D \cdot D_* &= WS_\Omega W_* + WS_{\varphi\Omega} + S_{\Omega\varphi} W_* + S_\varphi, \\
\Gamma \cdot \Gamma_* &= R, \\
G_0 &= \Gamma \cdot G \cdot D, \\
T &= \Gamma \cdot H \cdot (S_\Omega W_* + S_{\varphi\Omega}) D_*^{-1},
\end{aligned}
\tag{6.10}
$$

and substituting power spectral density (6.9) into (6.8), first variation of the performance criterion (6.8) with respect to the unknown filter-related function G_0 will become

$$
\delta J = \frac{1}{j} \int\limits_{-j\infty}^{j\infty} [(G_0 - T)\delta G_0 + \delta G_{0*}(G_{0*} - T_*)]ds.
\tag{6.11}
$$

Minimum of the performance criterion (6.8) is achieved when the first variation (6.11) turns to zero. Apparently, this is achieved when

$$
G = \Gamma^{-1}(T_0 + T_+)D^{-1}.
\tag{6.12}
$$

Here T_0 is the integer part of the function T, and T_+ is the part of the function T that contains only poles with negative real parts (stable poles) and is the result of the Wiener-Hopf separation procedure.

Using optimal solution (6.12) and assuming some spectral density for the angular rate and for the sensor noise, we can now derive corresponding static optimal filter transfer function.

The angular rate spectral must be selected which properly describes dynamics of the moveable object, where CVG is used. For most of the moveable objects, the following low-pass spectral density can be used:

$$
S_\Omega(s) = \frac{\sigma^2 B^2}{B^2 - s^2}.
\tag{6.13}
$$

Here B is the moveable object cut-off frequency, and σ is the standard deviation of the angular rate.

For the sake of simplicity, let us now neglect the cross-spectral densities between angular rate and noise ($S_{\varphi\Omega}(s) = S_{\Omega\varphi}(s) = 0$).

If sensor noise is a white noise and its spectral density is given by (6.3), the optimal filter can be calculated using (6.12) and will have the following transfer function

$$G(s) = [B\sqrt{1+\gamma^2}(s+\zeta k)]/[\gamma s^2$$
$$+ s\sqrt{\gamma(B^2\gamma + \zeta^2 k^2\gamma + 2\zeta kB\sqrt{1+\gamma^2})} \qquad (6.14)$$
$$+ \zeta kB\sqrt{1+\gamma^2}].$$

Transfer function (6.14) here has been also renormalized to remove steady-state error.

If the sensor noise is assumed to have high-pass spectral density (6.5), then the optimal filter transfer function is

$$G(s) = \frac{B(s+\zeta k)}{s^2\gamma + s\sqrt{\zeta k\gamma(2B + \zeta k\gamma)} + B\zeta k}. \qquad (6.15)$$

Depending on which of the senor noise model is found to be the most appropriate, either filter (6.14) or filter (6.15) can be used.

One should note that the obtained optimal filters (6.14) and (6.15) are *static*, and being expressed in terms of transfer functions can be easily implemented using simple analogue electronics at a low-level integrated circuitry. Contrary to static filtering, using filtering based on Kalman filter algorithm requires a microprocessor, which might not be feasible as in terms of cost efficiency as well as not providing sufficiently small high sampling frequency.

Let us now study the performances of the obtained optimal sensor noise filters by means of numerical simulations of the realistic CVG. Input angular rate is assumed to be in a form of square pulses. Results of numerical simulations of the "white" sensor noise filtering are shown in the Fig. 6.3.

These simulations were carried out for the $\gamma = 0.1$ and bandwidth of the angular rate $B = 3$ Hz. One should observe the good performance of the synthesised filter which efficiently removes the white sensor noise.

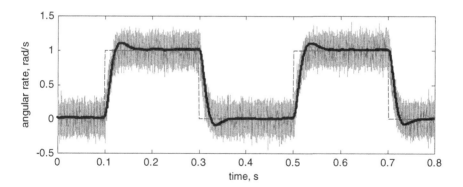

Fig. 6.3 Sensor noise filtering simulation (*dashed* input angular rate, *grey* noised output, *black* filtered output)

6.3 Process Noise Optimal Filtering

While sensor noise can be effectively removed by means of both static and dynamic optimal filtering, removing process noise, or rate-like disturbances, is somewhat more difficult. Disturbances are acting on CVG in the same way as the unknown angular rate, which makes them essentially undistinguishable from it (see Fig. 6.4).

Here in Fig. 6.4, $W(s)$ is the system transfer function given by (6.1), ψ is the process noise (stochastic disturbance), Ω_0 is the input angular rate, $G(s)$ is the optimal filter yet to be developed, Ω is the filtered output of the system, which in an ideal case is equal to the angular rate Ω_0.

One way to separate output resulting from the angular rate, from the output generated by the disturbances ψ is to take into account additional information about both angular rate and disturbances, and its spectral densities.

Let us assume, for example, that CVG is installed on a moveable object, such as aircraft or land vehicle, and power spectral density of the angular rate is given by (6.13). Process noise can be assumed to be a white noise with the spectral density (6.3) or a high-pass stochastic process described by (6.6). Apparently, the latter means better separation of the disturbances from the angular rate.

In order to synthesise process noise optimal filter, we can use the same Wiener approach described by (6.8)–(6.12). Sensor noise φ, as shown in Fig. 6.2, is the process noise ψ transformed by the CVG transfer function:

$$\varphi(s) = W(s) \cdot \psi(s). \tag{6.16}$$

Power spectral density of the sensor noise given by (6.16) is calculated using Wiener–Khinchin theorem as follows. For the white noise like disturbances (6.3) it is given by

$$S_{\varphi\varphi}(s) = |W_\Omega(s)|^2 S_\psi(s) = \frac{\gamma^2 \sigma^2 k^2 \zeta^2}{-s^2 + k^2\zeta^2}, \tag{6.17}$$

and for the case of high-pass disturbances (6.6) it is

$$S_{\varphi\varphi}(s) = \frac{\gamma^2 \sigma^2 k^2 \zeta^2 s^2}{(-s^2 + k^2\zeta^2)(-s^2 + B^2)}. \tag{6.18}$$

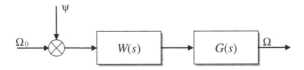

Fig. 6.4 CVG with added "rate-like" disturbances

Spectral densities (6.17) and (6.18) along with the suggested angular rate spectral density (6.13) can now be used to derive optimal filters based on the described above Wiener approach. After performing transformations according to (6.10), the optimal filters for the white noise disturbances is

$$G(s) = \frac{B\sqrt{1+\gamma^2}(s+\zeta k)}{\zeta k(\gamma s + B\sqrt{1+\gamma^2})},$$ (6.19)

and for the high-pass disturbances is given by

$$G(s) = \frac{B(s+\zeta k)}{\zeta k(B+\gamma s)}.$$ (6.20)

Depending on which of the disturbance model is found to be the most appropriate, either filter (6.19) or filter (6.20) should be used.

Results of numerical simulations of the filter (6.19) in case of the white noise disturbances and constant angular rate are shown in the Fig. 6.5.

These simulations are performed for the high-level disturbances ($\gamma = 1$) and low bandwidth of the angular rate ($B = 0.5$ Hz). When the bandwidth of the angular rate is increased, the disturbances filtering efficiency degrades.

Let us have a look at the efficiency of filtering as a function of the angular rate bandwidth B and the noise-to-rate ratio γ. These dependencies are shown in the Figs. 6.6 and 6.7.

Here, solid lines correspond to $\gamma = 1$ and the dotted lines to $\gamma = 0.5$. The lower the level of the standard deviation ratio σ/σ_0, the better filtering quality. One should note that when the standard deviation ratio is higher than one, filtering does not improve the quality of the angular rate measurements.

At the same time, while the bandwidth of the angular rate is lower than the bandwidth of CVG, the filters still can improve the characteristics of the sensors.

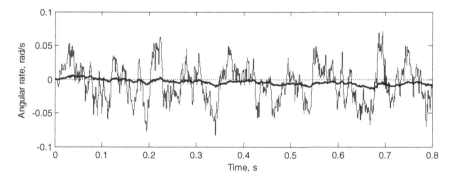

Fig. 6.5 Disturbances filtering simulations (*thin* unfiltered, *thick* filtered)

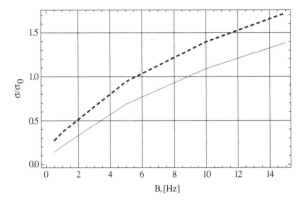

Fig. 6.6 "White" disturbances filtering efficiency

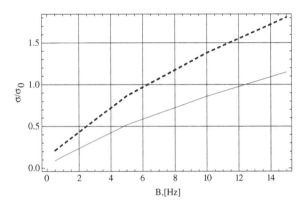

Fig. 6.7 "High-pass" disturbances filtering efficiency

6.4   Optimal Kalman Filter Synthesis

Despite the excellent performance of the Wiener filter in case of the stationary stochastic noises and disturbances, non-stationary noises still would require us to use adaptive Kalman filtering along with the corresponding computational hardware. Let us now demonstrate how to synthesise adaptive Kalman filter using demodulated dynamics of CVG.

In order to implement Kalman filter, we have to derive the difference model of the CVG dynamics in the following form:

$$\begin{cases} X_n = F \cdot X_{n-1} + w_{n-1}, \\ Z_n = C \cdot X_n + v_n. \end{cases} \tag{6.21}$$

Here X_n is the sampled state vector $X = \{\Omega \quad \Omega_0\}'$, Z_n is the measured state vector, $C = \begin{bmatrix} 1 & 0 \end{bmatrix}$ is the state measurement matrix, w_n and v_n are the process and sensor noises, respectively, and F is the state transition matrix, which can be obtained using the well-known dependency

$$F = L^{-1}\{(I \cdot s - A)^{-1}\}.$$ (6.22)

Here L^{-1} is the inverse Laplace transformation, A is the system matrix, which in case of the simplified system representation (6.1) is given as

$$A = \begin{bmatrix} -k\zeta & k\zeta \\ 0 & 0 \end{bmatrix}.$$ (6.23)

Note, that while deriving (6.23), the input angular rate Ω_0 was assumed to be a random walk (integrated white noise), which is the reason of zeroes in the second row of the matrix A. Substituting (6.23) into (6.22) results in

$$F = \begin{bmatrix} e^{-k\zeta t} & 1 - e^{-k\zeta t} \\ 0 & 1 \end{bmatrix}.$$ (6.24)

In order to verify state observability for the simplified model (6.23), let us calculate observability matrix as

$$Qo = \begin{bmatrix} C \\ C \cdot F \end{bmatrix} = \begin{bmatrix} 1 & 0 \\ e^{-k\zeta t} & 1 - e^{-k\zeta t} \end{bmatrix}.$$ (6.25)

Observability matrix (6.25) has full rank equal to 2, which satisfies the condition for state observability.

Governing equations for the discrete Kalman filter are as follows. Estimation of the system state X_n^- and error covariance matrix P_n^- are predicted as

$$\begin{aligned} X_n^- &= F \cdot \hat{X}_{n-1}, \\ P_n^- &= F \cdot \hat{P}_{n-1} \cdot F' + Q. \end{aligned}$$ (6.26)

where Q is the process noise w_n covariance. Next we calculate Kalman gain K_n, the corrected estimations of the system state $\hat{X}_n$, and the error covariance matrix $\hat{P}_n$ using the following expressions:

$$\begin{aligned} K_n &= P_n^- \cdot H' \cdot (H \cdot P_n^- \cdot H' + R)^{-1}, \\ X_n &= X_n^- + K_n \cdot (Z_n - H \cdot X_n^-), \\ \hat{P}_n &= (I - K_n \cdot H) \cdot P_n^-. \end{aligned}$$ (6.27)

Here R is the sensor noise v_n covariance. Calculated using (6.27) estimations of the system state and error covariance matrix are then used in (6.26) to make their next step prediction.

In order to verify Kalman filter performance, the same realistic CVG dynamics simulation is used, as in the previous simulations, but the Kalman filter block from the Signal Processing Blockset (Simulink/Matlab) is attached to the already demodulated output rate. Input angular rate has the shape of squared pulses with 1 rad/s amplitude. White noise is added to the output rate prior to being fed to the Kalman filter block. The results of the numerical simulations are shown in Figs. 6.8 and 6.9.

The following parameters of the CVG were used in simulations: $k = 500\,\text{Hz}$, $\zeta = 0.025$. Zero initial conditions were chosen for the state vector and the identity

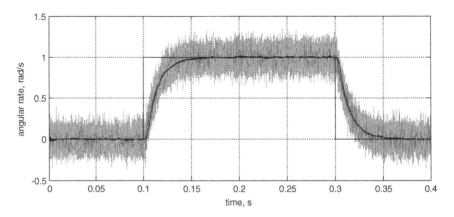

Fig. 6.8 Angular rate measurements (*grey* noised output, *dotted* actual output without noise, *solid* output estimation)

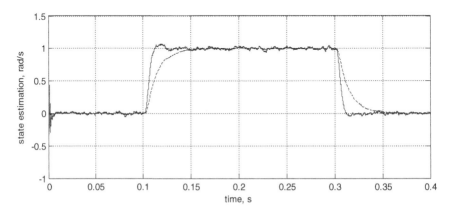

Fig. 6.9 State estimations over time (*solid* input angular rate, *dashed* output angular rate)

matrix has been used as an initial for the error covariance. Other parameters of the filter are as follows:

$$Q = \begin{bmatrix} 0 & 0 \\ 0 & 2 \cdot 10^{-6} \end{bmatrix}, \quad R = 0.01.$$

Analysing the graphs in Figs. 6.8 and 6.9, one can see that the added sensor noise has been successfully removed from the output, while the input angular rate has been estimated with some errors, however closer to the actual square pulse shape than measured output.

Let us now consider the case that angular rate is produced by the moving vehicle. In this case it can be modelled by means of a low-pass system described by the following equation:

$$\dot{\Omega} = -B \cdot \Omega + B \cdot \delta, \tag{6.28}$$

where B is the vehicle bandwidth, δ is the white noise. System matrix (6.23) and corresponding transition matrix (6.24) now become

$$A = \begin{bmatrix} -k\zeta & k\zeta \\ 0 & -B \end{bmatrix}, \tag{6.29}$$

and

$$F = \begin{bmatrix} e^{-k\zeta t} & \frac{e^{-k\zeta t} - e^{-Bt}}{B - k\zeta} k\zeta \\ 0 & e^{-Bt} \end{bmatrix} \tag{6.30}$$

correspondingly. One should note, that if $B = 0$ then matrices (6.29) and (6.30) become the ones from the previous model.

Simulation results for the state estimations of the low-pass angular rate case are shown in Fig. 6.10.

Sensor noise covariance matrices were taken the same with the previous case, and bandwidth has been chosen $B = 1$ Hz .

From the graph in Fig. 6.10 one can see that introducing bandwidth of the angular rate does not deliver any essential improvements to the quality of the input estimation. Moreover, as extensive analysis has demonstrated, increasing bandwidth introduces steady-state errors to the input angular rate estimation.

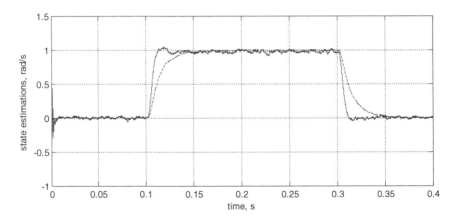

Fig. 6.10 State estimations for the low-pass angular rate (*solid* input angular rate, *dashed* output angular rate)

6.5 Cross-Coupling Compensation

An important performance parameter for a vibratory gyroscope is its zero-rate output or zero bias. Geometrical imperfections in the vibrating mechanical structure and/or the sense and drive electrodes as well as electrical coupling between these electrodes can cause an output signal in the absence of rotation. Specifically for shell gyroscope designs, cross-damping of the vibrating structure can also cause a rate-like output when there is no rotation.

In view of these problems, development of the efficient decoupling system for Coriolis vibratory gyroscopes is highly desired.

In order to solve the problem of the undesired cross-coupling compensation for CVGs, we have to determine the structure of the decoupling loop and identify its transfer functions.

6.5.1 Coupled Motion Equations and System Structure

In the most generalised form, motion equations of the CVG-sensitive element were previously derived in the form (6.29) as:

$$\begin{cases} \ddot{x}_1 + 2\zeta_1\omega_1\dot{x}_1 + (\omega_1^2 - d_1\Omega^2)x_1 = q_1 - g_1\Omega\dot{x}_2 - d_3\dot{\Omega}x_2, \\ \ddot{x}_2 + 2\zeta_2\omega_2\dot{x}_2 + (\omega_2^2 - d_2\Omega^2)x_2 = q_2 + g_2\Omega\dot{x}_1 + d_4\dot{\Omega}x_1. \end{cases}$$

Both equations here are solely coupled by the angular rate Ω, which results in the fundamental capability of such a system to measure external rotation. However, in a more realistic system, other cross-coupling will be present, manifesting itself as a cross-damping and cross-stiffness. Assuming small and quasi-constant angular rate ($\Omega^2 \approx 0$ and $\dot{\Omega} \approx 0$), incorporating cross-coupling terms, sensitive element motion equations will then be

$$\begin{cases} \ddot{x}_1 + 2\zeta_1\omega_1\dot{x}_1 + \omega_1^2 x_1 = q_1(t) - g_1\Omega\dot{x}_2 - d_{12}\dot{x}_2 - c_{12}x_2, \\ \ddot{x}_2 + 2\zeta_2\omega_2\dot{x}_2 + \omega_2^2 x_2 = q_2(t) + g_2\Omega\dot{x}_1 + d_{21}\dot{x}_1 + c_{21}x_1, \end{cases} \tag{6.31}$$

where d_{12} and d_{21} are the undesired cross-damping coefficients, c_{12} and c_{21} are the undesired cross-stiffness coefficients. These cross-coupling coefficients must be compensated, while the crucial term with the angular rate must be preserved.

By applying Laplace transformation to both sides of the system (6.31) with respect to zero initial conditions, we can obtain

$$\begin{cases} (s^2 + 2\zeta_1\omega_1 s + \omega_1^2)x_1(s) = q_1(s) - (g_1\Omega s + d_{12}s + c_{12})x_2(s), \\ (s^2 + 2\zeta_2\omega_2 s + \omega_2^2)x_2(s) = q_2(s) + (g_2\Omega s + d_{21}s + c_{21})x_1(s). \end{cases} \tag{6.32}$$

The sensitive element of CVG as an element of control systems and governed by the Eq. (6.32) can be represented by means of the structural scheme shown in Fig. 6.11.

Transfer functions in Fig. 6.11 are defined as

$$\begin{aligned} W_1(s) &= \frac{1}{s^2 + 2\zeta_1\omega_1 s + \omega_1^2}, \\ W_2(s) &= \frac{1}{s^2 + 2\zeta_2\omega_2 s + \omega_2^2}, \\ C_1(s) &= (g_1\Omega + d_{12})s + c_{12}, \\ C_2(s) &= (g_2\Omega + d_{21})s + c_{21}. \end{aligned} \tag{6.33}$$

Fig. 6.11 Structural diagram of coupled CVG

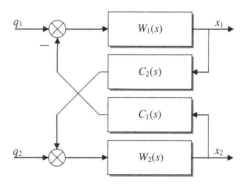

For this system its outputs can be found from the following system of algebraic equations:

$$\begin{cases} x_1(s) = W_1(s)[q_1(s) - C_1(s)x_2(s)], \\ x_2(s) = W_2(s)[q_2(s) + C_2(s)x_1(s)]. \end{cases} \tag{6.34}$$

Thus, solving system (6.34) and omitting Laplace variable "s", one can find outputs of CVG as:

$$\begin{aligned} x_1 &= \frac{W_1}{1 + C_1C_2W_1W_2}q_1 - \frac{C_1W_1W_2}{1 + C_1C_2W_1W_2}q_2, \\ x_2 &= \frac{W_2}{1 + C_1C_2W_1W_2}q_2 + \frac{C_2W_1W_2}{1 + C_1C_2W_1W_2}q_1. \end{aligned} \tag{6.35}$$

One should note, that in an ideal case of only useful Coriolis cross-coupling are present in the system

$$\begin{aligned} C_1(s) &\rightarrow C_{10}(s) = g_1\Omega s, \\ C_2(s) &\rightarrow C_{20}(s) = g_2\Omega s. \end{aligned} \tag{6.36}$$

Corresponding ideal system outputs can be obtained by substituting expressions (6.36) into expressions (6.35).

6.5.2 Decoupling System Synthesis

Let us consider the structure shown in Fig. 6.12 that is added to the outputs of the CVG-sensitive element as shown in Fig. 6.11.

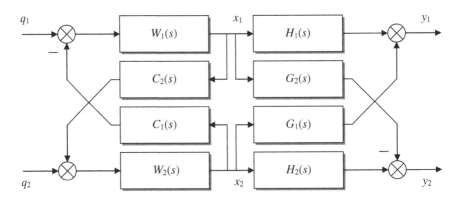

Fig. 6.12 CVG with added decoupling structure

Here transfer functions H_1, H_2, G_1 and G_2 are unknown and yet to be determined. Outputs of this system can be calculated as

$$
\begin{aligned}
y_1(s) &= H_1(s)x_1(s) + G_1(s)x_2(s), \\
y_2(s) &= H_2(s)x_2(s) - G_2(s)x_1(s).
\end{aligned}
\tag{6.37}
$$

Substituting (6.35) into (6.37) yields

$$
\begin{aligned}
y_1 &= \frac{H_1 W_1 + C_2 G_1 W_1 W_2}{1 + C_1 C_2 W_1 W_2} q_1 + \frac{G_1 W_2 - C_1 H_1 W_1 W_2}{1 + C_1 C_2 W_1 W_2} q_2, \\
y_2 &= \frac{H_2 W_2 + C_1 G_2 W_1 W_2}{1 + C_1 C_2 W_1 W_2} q_2 - \frac{G_2 W_1 - C_2 H_2 W_1 W_2}{1 + C_1 C_2 W_1 W_2} q_1.
\end{aligned}
\tag{6.38}
$$

Assuming that outputs (6.38) after the decoupling system must be identical to ideal system output, and comparing the corresponding transfer functions (coefficients in front of q_1 and q_2), the following system of equations can be produced:

$$
\begin{cases}
\dfrac{H_1 W_1 + C_2 G_1 W_1 W_2}{1 + C_1 C_2 W_1 W_2} = \dfrac{W_1}{1 + C_{10} C_{20} W_1 W_2}, \\[2mm]
\dfrac{G_1 W_2 - C_1 H_1 W_1 W_2}{1 + C_1 C_2 W_1 W_2} = -\dfrac{C_{10} W_1 W_2}{1 + C_{10} C_{20} W_1 W_2}, \\[2mm]
\dfrac{H_2 W_2 + C_1 G_2 W_1 W_2}{1 + C_1 C_2 W_1 W_2} = \dfrac{W_2}{1 + C_{10} C_{20} W_1 W_2}, \\[2mm]
-\dfrac{G_2 W_1 - C_2 H_2 W_1 W_2}{1 + C_1 C_2 W_1 W_2} = \dfrac{C_{20} W_1 W_2}{1 + C_{10} C_{20} W_1 W_2}.
\end{cases}
\tag{6.39}
$$

System (6.39) can now be solved for the unknown transfer functions H_1, H_2, G_1 and G_2, resulting in

$$
\begin{aligned}
G_1 &= \frac{(C_1 - C_{10}) W_1}{1 + C_{10} C_{20} W_1 W_2}, \\
G_2 &= \frac{(C_2 - C_{20}) W_2}{1 + C_{10} C_{20} W_1 W_2}, \\
H_1 &= \frac{1 + C_{10} C_2 W_1 W_2}{1 + C_{10} C_{20} W_1 W_2}, \\
H_2 &= \frac{1 + C_{20} C_1 W_1 W_2}{1 + C_{10} C_{20} W_1 W_2}.
\end{aligned}
\tag{6.40}
$$

Finally, substituting expressions (6.33) and (6.36) into solutions (6.40) results in the CVG decoupling system transfer functions:

$$G_1(s) = \frac{(c_{12} + d_{12}s)(s^2 + 2\zeta_2\omega_2 s + \omega_2^2)}{\Delta(s)},$$

$$G_2(s) = \frac{(c_{21} + d_{21}s)(s^2 + 2\zeta_1\omega_1 s + \omega_1^2)}{\Delta(s)},$$

$$H_1(s) = 1 + \frac{g_1 s(c_{21} + d_{21}s)}{\Delta(s)}\Omega,$$
(6.41)

$$H_2(s) = 1 + \frac{g_2 s(c_{12} + d_{12}s)}{\Delta(s)}\Omega.$$

Here the denominator is given by the following expression:

$$\Delta(s) = (s^2 + 2\zeta_1\omega_1 s + \omega_1^2)(s^2 + 2\zeta_2\omega_2 s + \omega_2^2) + g_1 g_2 s^2 \Omega^2.$$

Since it has been already assumed that angular rate Ω is small (e.g. $\Omega^2 \approx 0$), expressions (6.41) can be further simplified as

$$G_1(s) = \frac{c_{12} + d_{12}s}{s^2 + 2\zeta_1\omega_1 s + \omega_1^2},$$

$$G_2(s) = \frac{c_{21} + d_{21}s}{s^2 + 2\zeta_2\omega_2 s + \omega_2^2},$$

$$H_1(s) = 1 + \frac{g_1 s(c_{21} + d_{21}s)}{(s^2 + 2\zeta_1\omega_1 s + \omega_1^2)(s^2 + 2\zeta_2\omega_2 s + \omega_2^2)}\Omega,$$
(6.42)

$$H_2(s) = 1 + \frac{g_2 s(c_{12} + d_{12}s)}{(s^2 + 2\zeta_1\omega_1 s + \omega_1^2)(s^2 + 2\zeta_2\omega_2 s + \omega_2^2)}\Omega.$$

Analysis of the expressions (6.41) and (6.42) reveals the well-expected fact, that if all undesired cross-couplings, such as damping and stiffness, are absent, then these transfer functions are reduced to $G_1 = G_2 = 0$, $H_1 = H_2 = 1$. However, one could also note, that these transfer functions depend on the unknown angular rate Ω, which requires some additional steps to be undertaken to make the decoupling system viable.

6.5.3 Partial Decoupling System

Taking into consideration the fact that secondary oscillations are usually significantly smaller than primary oscillations, it is justifiable to assume that the influence of secondary oscillations on primary is negligibly small. Besides, only secondary oscillations output is used to measure angular rate. Hence, decoupling system can be simplified as shown in Fig. 6.13.

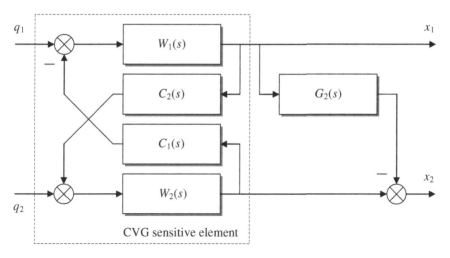

Fig. 6.13 CVG with the partial decoupling system

Here transfer function G_2 is given by (6.42). An important feature of the system shown in Fig. 6.13 is that it does not depend on the unknown angular rate.

Let us evaluate the performance of this system by means of numerical simulation. Measured angular rate with and without partial decoupling system is shown in Fig. 6.14.

Here $d_{12} = -d_{21} = 0.5$, $c_{12} = c_{21} = 50,000$. One can see that even partial decoupling system significantly improves the performance of CVG.

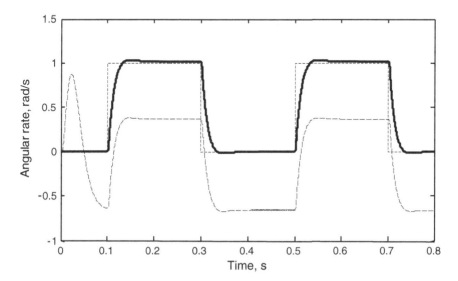

Fig. 6.14 Results of numerical simulations (*dotted* input angular rate, *dashed* without decoupling, *solid* with decoupling)

6.5.4 Linearised Complete Decoupling System

Although the above-considered partial decoupling system somewhat decouples CVG from undesired couplings, it still might not be sufficient for the high-performance devices. At the same time, complete decoupling system based upon transfer functions (6.41) is not feasible due to the presence of the unknown angular rate as a coefficient in its transfer functions. However, in the linearised expressions (6.42), angular rate is present only as an additional input to the decoupling system. Such peculiarity allows us to build the decoupling system by means of feeding decoupled angular rate back to the decoupling system, as shown in Fig. 6.15.

Here, in Fig. 6.15, block "CVG Sensitive Element" has been already shown in Fig. 6.11, block "Secondary Demodulator" demodulates secondary oscillations and produces measured angular rate as its output. Finally, feedback transfer function $H_{20}(s)$ is given by the following expression:

$$H_{20}(s) = \frac{g_2 s(c_{12} + d_{12}s)}{(s^2 + 2\zeta_1\omega_1 s + \omega_1^2)(s^2 + 2\zeta_2\omega_2 s + \omega_2^2)}. \tag{6.43}$$

Expression (6.43) is derived directly from (6.42) as a coefficient to the angular rate in the expression for the transfer function $H_2(s)$.

In order to evaluate the performance of the linearised decoupling system in comparison with the partial decoupling system, let us numerically simulate the operation of a realistic CVG with these two systems. The essential part of a unit step transient process is shown in Fig. 6.16.

Although the system presented in Fig. 6.15 is nonlinear, its performance is apparently better than the performance of the partial decoupling system.

Studied here, the decoupling system, its structure and transfer functions allows considerable improvement of the CVG performances by means of practical elimination of undesired cross-couplings. Partial decoupling system is suggested for the

Fig. 6.15 Linearised complete decoupling system

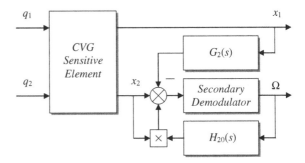

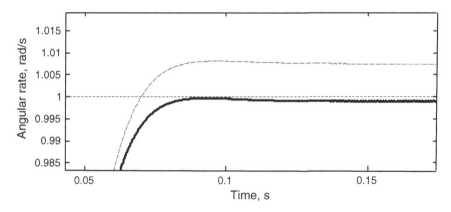

Fig. 6.16 Decoupling systems performance comparison (*dotted* input angular rate, *dashed* partial decoupler, *solid* linearised complete decoupler)

devices with low accuracy requirements due to its simplicity. For the high-performance sensors, usage of the linearised complete decoupling system can be justified. Both these approaches improve bias and scale factor stability of CVG.

6.6   Temperature Errors Compensation

One of the well-recognised major sources of bias instabilities in CVGs is temperature variations. It affects practically all performances of CVGs. The following figures demonstrate significant temperature-related zero-rate output that has been observed during experimental tests of CVG. For the temperature profile, shown in Fig. 6.17, and zero angular rate, uncompensated CVG output is shown in Fig. 6.18.

It is believed that the temperature variations cause this bias through the temperature-dependent cross-damping. In this case, the excited primary oscillations of the sensitive element are capable to induce secondary (output) oscillations even without external rotation being applied to the sensor.

6.6.1   Cross-Damping Transfer Function

Let us have a look at the CVG-sensitive element motion equations when temperature-related cross-damping is present in the system:

$$\begin{cases} \ddot{x}_1 + 2\zeta_1\omega_1\dot{x}_1 + (\omega_1^2 - d_1\Omega^2)x_1 + (g_1\Omega + 2\zeta_{21}\omega_2)\dot{x}_2 + d_3\dot{\Omega}x_2 = q_1(t), \\ \ddot{x}_2 + 2\zeta_2\omega_2\dot{x}_2 + (\omega_2^2 - d_2\Omega^2)x_2 - (g_2\Omega + 2\zeta_{12}\omega_1)\dot{x}_1 - \dot{\Omega}x_1 = q_2(t). \end{cases} \qquad (6.44)$$

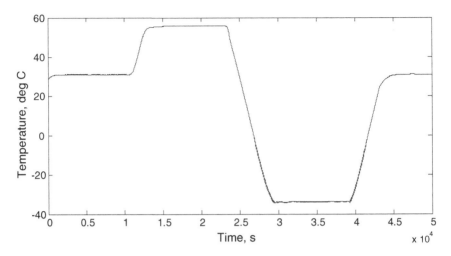

Fig. 6.17 Temperature profile

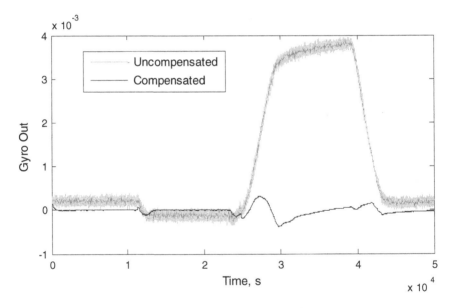

Fig. 6.18 CVG output with and without temperature compensation

Here ζ_{12} and ζ_{21} are the relative cross-damping coefficients. Indeed, even if there is no external rotation ($\Omega = 0$), secondary oscillations linearly related to ζ_{12} will still be registered.

Constant cross-coupling through the damping can be removed by calibration. However, calibration is unable to deal effectively with the varying in time damping ζ_{12} due to the temperature variations.

Let us now derive mathematical model for the temperature caused output. Following the demonstrated in Chap. 4 method of analysing CVG-sensitive element dynamics in demodulated signals, we can represent complex secondary amplitude in Laplace domain as:

$$A_2(s) = W_2^\Omega(s) \cdot \Omega(s) + W_2^\zeta(s) \cdot \zeta_{12}(s). \tag{6.45}$$

Here transfer function are given by the following expressions:

$$\begin{aligned}
W_2^\Omega(s) &= \frac{q_{10}(j\omega g_2 + d_4 s)}{[(s + j\omega)^2 + 2\zeta_2\omega_2(s + j\omega) + \omega_2^2](\omega_1^2 - \omega^2 + 2j\omega_1\zeta_1\omega)}, \\
W_2^\zeta(s) &= \frac{2j\omega\omega_2}{[(s + j\omega)^2 + 2\zeta_2\omega_2(s + j\omega) + \omega_2^2](\omega_1^2 - \omega^2 + 2j\omega_1\zeta_1\omega)}.
\end{aligned} \tag{6.46}$$

It is important to remember that part of the secondary amplitude due to the cross-damping will be undistinguishable from the one caused by the angular rate. Let us, therefore, derive the transfer function relating input cross-damping to the output angular rate as

$$\Omega^\zeta(s) = W_\Omega^\zeta(s) \cdot \zeta_{12}(s), \tag{6.47}$$

where $\Omega^\zeta(s)$ is the measured erroneous angular rate caused by the cross-damping. Apparently, transfer function $W_\Omega^\zeta(s)$ can be expressed using transfer functions from (6.46) as

$$W_\Omega^\zeta(s) = \frac{W_2^\zeta(s)}{W_2^\Omega(s \to 0)} = \frac{2\omega_2(\omega_2^2 - \omega^2 + 2j\omega_2\omega\zeta_2)}{g_2(\omega_2^2 - \omega^2 + 2\omega_2\zeta_2 s + 2j\omega(s + \omega_2\zeta_2))}. \tag{6.48}$$

Transfer function (6.48) can be further simplified using assumptions, similar to those used in Chap. 4. Namely, we can assume that the natural frequencies are equal ($\omega_1 = \omega_2 = k$) as well as relative damping coefficients ($\zeta_1 = \zeta_2 = \zeta$), and primary oscillations excitation frequency is $\omega = k\sqrt{1 - 2\zeta^2}$. With these assumptions, the transfer function (6.48) becomes

$$W_\Omega^\zeta(s) = \frac{2k^2\zeta}{g_2(s + k\zeta)}. \tag{6.49}$$

The transfer function (6.49) allows efficient analysis of errors due to the cross-damping, which not only is present in the system, but can vary due to different reasons.

6.6.2 Empirical Modelling of Cross-Damping

Assuming that the cross-damping coefficient is a function of the temperature shift T from the calibration temperature, it can be approximated using polynomial as

$$\zeta_{12} = \zeta_{12}(T) \approx \sum_{i=0}^{n} \zeta_i^T T^i. \tag{6.50}$$

Temperature-related coefficients ζ_i^T can be determined experimentally when ambient temperature is known (measured) and angular rate is absent (see the Figs. 6.17 and 6.18). However, in most of the cases, we observe angular rate as the gyro output. In order to relate angular rate to the input cross-damping, let us use the steady state of the transfer function (6.49) as

$$\Omega(T) = W_{\Omega}^{\zeta}(s \to 0)\zeta_{12}(T) \approx \frac{2k}{g_2} \sum_{i=0}^{n} \zeta_i^T T^i = \sum_{i=0}^{n} \Omega_i^T T^i. \tag{6.51}$$

Parameters Ω_i^T of the cross-damping model (6.51) can now be identified from the experimental data and found to have the following values: $\Omega_0^T = 10.0792 \times 10^{-3}$, $\Omega_1^T = -40.631 \times 10^{-5}$, $\Omega_2^T = 70.7044 \times 10^{-7}$, $\Omega_3^T = -50.8598 \times 10^{-9}$.

Influence of the higher order components found to be negligible. In order to validate the cross-damping model (6.51), obtained temperature-related angular rate can be subtracted from the gyroscope output, producing compensated output as shown in Fig. 6.18 (compensated line). As one can see, model (6.51) successfully compensates bias for the steady temperature, while performs only fair during temperature transitions.

6.6.3 Temperature Compensation System

In order to deal successfully with the transient processes in CVG dynamics due to temperature, let us synthesise the temperature compensation system using the cross-coupling compensation technique described in this chapter earlier in the previous section. The structure of a simple partial decoupling system is shown in Fig. 6.13. In case of temperature-related cross-damping compensation, transfer function $G_2(s)$ in the decoupling system is given by

$$G_2(s) = \frac{2\zeta_{12}\omega_1 s}{s^2 + 2\zeta_2\omega_2 s + \omega_2^2}. \tag{6.52}$$

Here ζ_{12} is the temperature-dependent cross-damping coefficient given by (6.50). By taking temperature measurements from the temperature sensor, one can now combine these readings with the measured primary oscillations to implement low-level (before demodulation) temperature compensation as shown in the Fig. 6.19.

Results of numerical simulations of this system operation are shown in Fig. 6.20.

In these numerical simulations, temperature has a sinusoidal shape ranging from -50 to $50\ ^{\circ}\mathrm{C}$ and period of 1 s. One can see that the proposed temperature

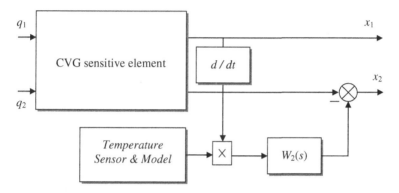

Fig. 6.19 Low-level temperature compensation system

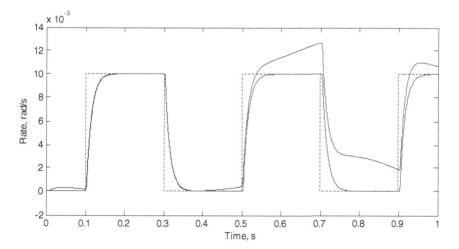

Fig. 6.20 Temperature compensation system simulation (*dashed line* input angular rate, *thin line* uncompensated output, *thick line* compensated output)

compensation system successfully removed the effect of cross-damping variations
due to temperature. However, this system still requires temperature sensor being
used in the system.

6.6.4 Optimal Filtering of Temperature Errors

Analysing (6.45) one could note that temperature influences the output of a CVG
exactly like an angular rate and the temperature-related output is undistinguishable
from the angular rate measurements:

$$\Omega(s) = W(s) \cdot \Omega_0(s) + W_\zeta(s) \cdot \zeta_{12}(s). \tag{6.53}$$

Here $W(s)$ is the system transfer function given by (6.1), and $W_\zeta(s)$ is the
cross-damping transfer function given by the expression (6.49). In this sense,
temperature influence could be treated as a process noise or disturbance to the CVG
system, as shown in Fig. 6.4. This allows us to derive optimal process noise filter
using the method, studied earlier, to produce a temperature error compensation
system that does not require temperature measurements, but uses statistical characteristics of temperature variations to minimise temperature errors.

By utilising similarities in transfer functions $W(s)$ and $W_\zeta(s)$, we can define
disturbances as

$$\psi(s) = \frac{2k}{g_2} \zeta_{12}(s). \tag{6.54}$$

Assuming that CVG is installed on a manoeuvrable object, such as an unmanned
aerial vehicle, for example, or land vehicle, its power spectral density can be
represented as a low-pass model, similar to the one used earlier:

$$S_\Omega(s) = \frac{\sigma^2 B^2}{B^2 - s^2}. \tag{6.55}$$

It is apparent that the temperature variations are slow and therefore could be
adequately represented by the following random walk model

$$S_\zeta(s) = \frac{\gamma^2 \sigma^2}{-s^2}. \tag{6.56}$$

Power spectral densities (6.55) and (6.56) can now be used to synthesise optimal
filter, reducing errors caused by the temperature variations.

In terms of the Wiener algorithm for calculating optimal process noise filters
studied in Sect. 6.3, and spectral density $S_{\varphi\varphi}(s)$ can be calculated from (6.55) using
Wiener–Khinchin theorem as follows:

$$S_{\varphi\varphi}(s) = |W_\Omega(s)|^2 S_\psi(s)$$

$$= |W_\Omega(s)|^2 \frac{4k^4}{g_2^2} S_\zeta(s) \qquad (6.57)$$

$$= \frac{4\gamma^2 \sigma^2 k^4 \zeta^2}{-s^2 g_2^2 (-s^2 + k^2 \zeta^2)} .$$

Spectral density (6.57) along with the spectral density (6.55) can now be used to derive optimal filters based on the formula (6.12). After performing all necessary transformations, the optimal filter is found in the following form:

$$G(s) = \frac{2Bk\gamma(k\zeta + s)}{k\zeta \left(2Bk\gamma + s\sqrt{g_2^2 B^2 + 4k^2\gamma^2}\right)} . \qquad (6.58)$$

Optimal temperature errors filter (6.58) can be used to reduce effect of the temperature variations on CVG performances. It is also important to note that filter (6.58) is a static transfer function and therefore does not require computational devices, and can be implemented using analogue electronics as an application-specific integrated circuit.

6.7 Whole-Angle Force Rebalance Control

The necessity to measure the angle of rotation instead of angular rate (whole-angle measurement mode) led to specifically designed sensitive elements, behaving similarly to Foucault pendulum. In this section, a different approach is considered when a conventional angular rate sensing CVG is provided with a feedback controller providing similar whole-angle operation while restraining the rate sensing output of the sensitive element.

In terms of the demodulated (envelope) signals, CVG along with the negative feedback loop can be represented as a control system shown in Fig. 6.21.

Here Ω—is the actual output of the gyroscope (measured angular rate), Ω_0—is the unknown angular rate (system input) and $W(s)$ is the CVG transfer function given by (6.1). The goal is to design such a feedback controller $G(s)$, producing output y, which is being modulated and applied to the secondary mode of the CVG-sensitive element will limit the actual output of the gyro. At the same time, signal y itself must represent integrated angular rate, e.g. angle of rotation.

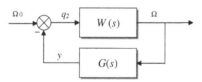

Fig. 6.21 CVG with the feedback controller

The transfer function from the input angular rate Ω_0 to the feedback output y is as follows:

$$W_y(s) = \frac{W(s) \cdot G(s)}{1 + W(s) \cdot G(s)}. \tag{6.59}$$

In order to perform the required task of the integrated angular rate measurement ("whole angle" mode), transfer function (6.59) must be equal to the simple integrator $1/s$, which results in the following equation:

$$\frac{W(s) \cdot G(s)}{1 + W(s) \cdot G(s)} = \frac{1}{s}. \tag{6.60}$$

Substituting (6.1) into the Eq. (6.60) and solving this equation for the unknown feedback transfer function $G(s)$ results in

$$G(s) = \frac{\zeta k + s}{\zeta k(s - 1)}. \tag{6.61}$$

It is important to remember that the transfer function (6.61) is derived in terms of the demodulated envelope signals. It means, that in order to apply its output as an actuation to the secondary mode of the sensitive element (acceleration q_2), it must be modulated with the differentiated output of the primary mode as follows to make it identical to the Coriolis force, acting along the secondary motion coordinate:

$$q_2(t) = g_2 \cdot y(t) \cdot \dot{x}_1(t). \tag{6.62}$$

Applying signal (6.62) to the sensitive element results in reduction of its displacements, while the feedback output y becomes the new output of the gyro, implementing the whole-angle operation.

Simulation schematics of CVG with the feedback controller is shown in Fig. 6.22.

Here subsystem "CVG dynamics" simulates sensitive element dynamics based on the complete generalised equations. Figure 6.23 demonstrates demodulated signals produced by the simulation: input angular rate (dashed), highly noised measured angular rate (grey) and the compensated actual output of the CVG.

One could note that the actual output of the gyro (solid line) is actually less than the input angular rate. At the same time, output of the feedback controller (shown in Fig. 6.24) produces integrated angular rate (angle of rotation) while reducing the effect from the noise in the feedback loop.

Presented here is the synthesis of a feedback controller result in a system, which being applied to a conventional CVG allows to implement the whole-angle force rebalance mode for the gyro. The obtained controller reduces the sensitive element deflections and the influence of the measurement noise on the output angle of rotation as well.

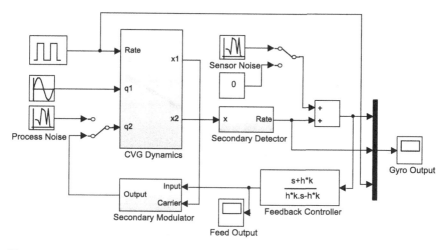

Fig. 6.22 Simulating CVG control operation in Simulink

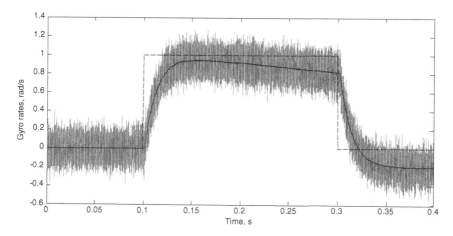

Fig. 6.23 CVG signals: *solid* secondary output in the force rebalance mode, *grey* noised secondary output, *dashed* input angular rate

Resume

One could always design the best possible sensitive element given the required performances. But it is almost impossible to fabricate the sensitive element to the same level of perfection at which it has been designed. Plenty of imperfections will certainly be introduced during fabrication, sometimes rendering to nothing all the advantages of the optimised design. Improving the fabrication process to a higher

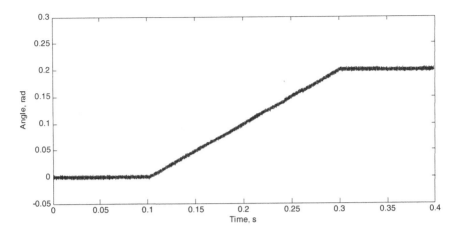

Fig. 6.24 Integrated (whole angle) CVG feedback output

level of tolerances is one way, but it significantly increases fabrication cost. Another way, and it is believed to be preferable, is to compensate the effect of imperfections by properly synthesised signal processing and control.

Further Reading

1. Savet P (1961) Gyroscopes: theory and design. McGraw-Hill, New York
2. Quick W (1964) Theory of vibrating string as an angular motion sensor. J Appl Mech 31(3):523–534
3. Friedland B, Hutton MF (1978) Theory and error analysis of vibrating-member gyroscope. IEEE Trans Autom Control 23:545–556
4. O'Connor J, Shupe D (1983) Vibrating beam rotation sensor. U.S. Patent 4381672
5. Ragan R (1984) Inertial technology for the future. IEEE Trans Aerosp Electron Syst V-AES20 (4):414–440
6. Boxenhorn B (1984) Planar inertial sensor. US Patent 4598585, 8 July 1986
7. Boxenhorn B (1988) A vibratory micromechanical gyroscope. In: Proceedings of the AIAA guidance and control conference, Minneapolis, p 1033
8. Boxenhorn B, Greiff P (1988) A vibratory micromechanical gyroscope. In: Proceedings of the AIAA guidance and controls conference Minneapolis, USA, pp 1033–1040
9. Tang W, Nguyen T-C, Howe R (1989) Laterally driven polysilicon resonant microstructures. In: Proceedings of the micro electro mechanical systems, 1989, An investigation of micro structures, sensors, actuators, machines and robots. IEEE, pp 53–59
10. Fujimura S, Yano K, Kumasaka T, Ariyoshi H, Ono O (1989) Vibration gyros and their applications. In: Proceedings of the IEEE international conference on consumer electronics, p 116–117
11. Buser R, De Rooij N (1989) Tuning forks in silicon. In: IEEE Micro-electro-mechanical systems workshop, Salt Lake City, USA, pp 94–95
12. Söderkvist J (1990) A mathematical analysis of flexural vibrations of piezoelectric beams with applications to angular rate sensors. Ph.D. thesis, Uppsala University, Sweden
13. Greiff P, Boxenhorn B, King T, Niles L (1991) Silicon monolithic gyroscope. In: Transducers '91, digest of technical papers, international conference on solid state sensors and actuators, pp 966–969
14. Söderkvist J (1991) Piezoelectric beams and vibrating angular rate sensors. In: IEEE transactions on ultrasonics, ferroelectrics and frequency control, vol 38, no 3, pp 271–280
15. Fujishima S, Nakamura T, Fujimoto K (1991) Piezoelectric vibratory gyroscope using flexural vibration of a triangular bar. In: Proceedings of the 45th annual symposium on frequency control, pp 261–265
16. Abe H, Yoshida T, Turuga K (1992) Piezoelectric-ceramic cylinder vibratory gyroscope. Japanese Journal of Applied Physics 31(Part 1, No. 9B):3061–3063
17. Lawrence A (1993) Modern inertial technology: navigation, guidance and control. Springer, New York
18. Bernstein J, Cho S, King A, Kourepenis A, Maciel P, Weinberg M (1993) A micromachined comb-drive tuning fork rate gyroscope. In: Proceedings of the micro electro mechanical systems, MEMS '93. An investigation of micro structures, sensors, actuators, machines and systems. IEEE, pp 143–148

19. Bernstein I, Weinberg M (1994) Comb drive micromechanical tuning fork gyro. US Patent 5349855, 27 Sept 1994
20. Burdess J, Harris A, Cruickshank J, Wood D, Cooper G (1994) A review of vibratory gyroscopes. J Eng Sci Educ V3(6):249–254
21. Söderkvist J (1994) Micromachined gyroscopes. Sens Actuators A 43(1–3):65–71
22. Putty M, Najafi K (1994) A micromachined vibrating ring gyroscope. In: Solid-state sensor and actuator workshop Hilton Head, pp 213–220
23. Lynch D (1995) Vibratory gyro analysis by the method of averaging. In: Proceedings of the 2nd St. Petersburg conference on gyroscopic technology and navigation, St. Petersburg, pp 26–34
24. Tanaka K, Mochida Y, Sugimoto M, Moriya K, Hasegawa T, Atsuchi K, Ohwada K (1995) A micromachined vibrating gyroscope. Sens Actuators A 50(1–2):111–115
25. Wood D, Cooper G, Burdess J, Harris A, Cruickshank J (1995) A silicon membrane gyroscope with electrostatic actuation and sensing. In: SPIE proceedings in microfabrication and micromachining, vol 2642, pp 74–83
26. Maenaka K, Fujita T, Konishi Y, Maeda M (1996) Analysis of a highly sensitive silicon gyroscope with cantilever beam as vibrating mass. Sens Actuators A 54(1–3):568–573
27. Greiff P, Antkowiak B, Campbell J, Petrovich A (1996) Vibrating wheel micromechanical gyro. In: Proceedings of the IEEE position location and navigation symposium, pp 31 –37
28. Clark W, Howe R, Horowitz R (1996) Micromachined Z-axis vibratory rate gyroscope. In: Technical digest of the solid-state sensor and actuator workshop, Hillon Head, South Carolina, pp 283–287
29. Paoletti F, Gretillat M-A, de Rooij N (1996) A silicon micromachined vibrating gyroscope with piezoresistive detection and electromagnetic excitation. In: Micro-electro-mechanical systems, MEMS '96, pp 162–167
30. Langmaid C (1996) Vibrating structure gyroscopes. Sens Rev 16(1):14–17
31. Clark W (1996) Micromachined vibratory rate gyroscopes. Ph.D. Dissertation, U.C. Berkeley
32. Titterton D, Weston J (1997) Strapdown inertial navigational technology. Peter Peregrinus Ltd, Lavenham
33. Niu M, Xue W, Wang X, Xie J, Yang G, Wang W (1997) Vibratory wheel gyroscope. In: Proceedings of the transducers'97, pp 891–894
34. Juneau T, Pisano A, Smith J (1997) Dual axis operation of a micromachined rate gyroscope. In: Proceedings of the international conference on solid state sensors and actuators, TRANSDUCERS '97 Chicago, vol 2, pp 883–886
35. Hulsing R (1998) MEMS inertial rate and acceleration sensor. IEEE AES Syst Mag 13:17–23
36. Yazdi N, Ayazi F, Najafi K (1998) Micromachined intertial sensors. Proc IEEE 86 (8):1640–1659
37. Solomon S (1998) Sensors handbook. McGraw Hill Handbooks, New York
38. Oh Y, Lee B, Baek S, Kim H, Kim J, Kang S, Song C (1998) A tunable vibratory microgyroscope. Sens Actuators A 64:51–56
39. Allen J, Kinney R, Sarsfield J, Daily M, Ellis J, Smith J, Montague S, Howe R, Horowitz R, Pisano A, Lemkin M, Clark W, Juneau T (1998) Integrated micro-electro-mechanical sensor development for inertial applications. IEEE AES Syst Mag, 36–40
40. Yachi M, Ishikawa H, Satoh Y, Takahashi Y, Kikuchi K (1998) Design Methodology of single crystal tuning fork gyroscope for automotive applications. In: Proceeding of the IEEE ultrasonics symposium, vol 1, pp 463–466
41. Geiger W, Folkmer B, Sobe U, Sandmaier H, Lang W (1998) New designs of micromachined vibrating rate gyroscopes with decoupled oscillation modes. Sens Actuators A 66(1–3):118–124
42. Kourepenis A, Bernstein J, Connelly J, Elliott R, Ward P, Weinberg M (1998) Performance of MEMS inertial sensors. In: IEEE symposium on position location and navigation, pp 1–8
43. Degani O, Seter D, Socher E, Kaldor S, Nemirovsky Y (1998) Optimal design and noise consideration of micromachined vibrating rate gyroscope with modulated integrative differential optical sensing. IEEE J Microelectromech Syst 7(3):329–338

44. Hopkin I, Fell C, Townsend K, Mason T (1999) Vibrating structure gyroscope. US Patent 5932804

45. Dong Y, Gao Z, Zhang R, Chen Z (1999) A vibrating wheel micromachined gyroscope for commercial and automotive applications. In: Proceedings of the 16th IEEE instrumentation and measurement technology conference, vol 3, pp 1750–1754

46. Shkel A, Horowitz R, Seshia A, Park S, Howe R (1999) Dynamics and control of micromachined gyroscopes. In: Proceedings of the American control conference, vol 3, pp 2119–2124

47. Leland R (2001) Mechanical thermal noise in vibrating gyroscopes. In: Proceedings of the American control conference, 25–27 June 2001, pp 3256–3261

48. Apostolyuk V, Logeeswaran VJ, Tay F (2002) Efficient design of micromechanical gyroscopes. J Micromech Microeng 12:948–954

49. Painter C, Shkel A (2003) Active structural error suppression in MEMS vibratory rate integrating gyroscopes. IEEE Sens J 3(5):595–606

50. Apostolyuk V, Tay F (2004) Dynamics of micromechanical coriolis vibratory gyroscopes. Sens Lett 2(3–4):252–259

51. Fraden J (2004) Handbook of modern sensors: physics, designs, and applications. Spinger, New York

52. Apostolyuk V (2006) Theory and design of micromechanical vibratory gyroscopes. In: Leondes CT (ed) MEMS/NEMS Handbook, Springer, vol 1, Chapter 6, pp 173–195

53. Weinberg M, Kourepenis A (2006) Error sources in in-plane silicon tuning-fork MEMS gyroscopes. J Microelectromech Syst 15(3):479–491

54. Saukoski M, Aaltonen L, Halonen K (2007) Zero-rate output and quadrature compensation in vibratory MEMS gyroscopes. IEEE Sens J 7(12):1639–1652

55. Dong Y, Kraft M, Hedenstierna N, Redman-White W (2008) Microgyroscope control system using a high-order band-pass continuous-time sigma-delta modulator. Sens Actuators A 145–146:299–305

56. Antonello R, Oboe R, Prandi L, Caminada C, Biganzoli F (2009) Open loop compensation of the quadrature error in MEMS vibrating gyroscopes. In: 35th annual conference of IEEE industrial electronics, IECON '09, pp 4034–4039

57. Jiancheng F, Jianli L (2009) Integrated model and compensation of thermal errors of silicon microelectromechanical gyroscope. IEEE Trans Instrum Measur 58(9):2923–2930

58. Antonello R, Oboe R (2011) MEMS gyroscopes for consumer and industrial applications. Microsensors, Intech, pp 253–280

59. Tatar E, Alper S, Akin T (2012) Quadrature-error compensation and corresponding effects on the performance of fully decoupled MEMS gyroscopes. J Microelectromech Syst 21(3):656–667

60. Prikhodko I, Zotov S, Trusov A, Shkel A (2012) Foucault pendulum on a chip: rate integrating silicon MEMS gyroscope. Sens Actuators A 177:67–78

Printed in the United States
By Bookmasters